The Politics of Nuclear Power

A History of the Shoreham Nuclear Power Plant

Technology, Risk, and Society

An International Series in Risk Analysis

VOLUME 5

The titles published in this series are listed at the end of this volume.

The Politics of Nuclear Power

A History of the Shoreham Nuclear Power Plant

By

David P. McCaffrey
Department of Public Administration and Policy,
and Organizational Studies Program,
University at Albany,
State University of New York, U.S.A.

KLUWER ACADEMIC PUBLISHERS
DORDRECHT / BOSTON / LONDON

Library of Congress Cataloging-in-Publication Data

McCaffrey, David P., 1930-
The politics of nuclear power : a history of the Shoreham Nuclear Power Plant / by David P. McCaffrey.
p. cm. -- (Technology, risk, and society ; v. 5)
Includes bibliographical references and index.
ISBN 0-7923-1035-7 (alk. paper)
1. Shoreham Nuclear Powerplant (N.Y.)--History. 2. Nuclear power plants--Government policy--United States--History. 3. Nuclear industry--Government policy--United States--History. 4. Electric utilities--Government policy--United States--History. 5. Energy policy--United States--History. I. Title. II. Series.
HD9698.U54S546 1990
333.792'4'0974727--dc20 90-19154

ISBN 0-7923-1035-7

Published by Kluwer Academic Publishers,
P.O. Box 17, 3300 AA Dordrecht, The Netherlands.

Kluwer Academic Publishers incorporates
the publishing programmes of
D. Reidel, Martinus Nijhoff, Dr W. Junk and MTP Press.

Sold and distributed in the U.S.A. and Canada
by Kluwer Academic Publishers,
101 Philip Drive, Norwell, MA 02061, U.S.A.

In all other countries, sold and distributed
by Kluwer Academic Publishers Group,
P.O. Box 322, 3300 AH Dordrecht, The Netherlands.

Printed on acid-free paper

Printed in the Netherlands

CONTENTS

Chapter 1

INTRODUCTION

Several individuals noted the potentially important civilian uses of atomic energy shortly after the bombings of Hiroshima and Nagasaki in 1945. That year J. Robert Oppenheimer told a national radio audience that "in the near future" it would be possible to generate profitable electric power from "controlled nuclear chain reaction units" (reactors). It was suggested that, after fifteen to twenty-five years of development, mature nuclear technology could provide virtually inexhaustible, cheap energy given the abundance of nuclear fuel. Admiral Lewis Strauss, the Chairman of the Atomic Energy Commission, stated that atomic power would generate electricity "too cheap to meter" (A statement that, according to Brookhaven National Laboratories' physicist Herbert Kouts, immediately "caused consternation among his technical advisors" [Kouts, 1983: 3]). For a brief period it was thought that airplanes would fly using atomic power, and homes would install small nuclear reactors for heat and hot water.

In the 1950s and early 1960s a small number of prototype nuclear reactors came on line in the United States. The first power plant protoype reactor began operation in Shippingport, Pennsylvania in 1957. It was followed by the Dresden 1 unit near Chicago in 1959, the Yankee plant in Rowe, Massachusetts (1960), and the Indian Point (New York) and Big Rock Point (Michigan) plants in 1962. These five plants had a combined generating capacity of less than 800 megawatts (800 MW), or less than one percent of the total American electricity generating capacity in 1962. The United States plants constituted less than half of the world's total nuclear electric power generating capacity at the time.

The amount of nuclear generated capacity increased rapidly in the 1960s and 1970s. By the end of 1975, 157 nuclear plants with a generating capacity of approximately 72,000 MW were operating in nineteen countries. Most nations which already had nuclear plants were building additional plants (the Netherlands and Pakistan were exceptions). In April 1975 an international conference on "The Maturity of Nuclear Power" was held in Paris. A.E. Hawkins, Chairman of the United Kingdom Electricity

Generating Board, called the theme "both commendable and appropriate." He said that there once had been doubts about the technical and economic viability of nuclear power. However, "'all of these questions had been resolved.' Nuclear power had indeed 'matured'" (Bupp and Derian, 1978: 6-7).

By December, 1976, a total of sixty reactors with an aggregate capacity of 43,000 MW were completed in the United States. An additional 146 were under construction or on order. In 1976 the total capacity of the nation's generating network was approximately 500,000 MW, and so nuclear generated electric power represented about nine percent of the nation's on-line capacity.

But the rapid growth of planned nuclear capacity virtually stopped by the late 1970s. Table 1 indicates the trend in orders for nuclear plants in the United States. Prior to 1972, 131 orders were placed and none were cancelled. From 1972 through 1982, 120 orders were placed and 100 were cancelled. Most striking is that from 1979 through 1982 no orders were placed, and thirty-three were cancelled. From 1983 through 1988 utilities took no new orders, thirteen more nuclear units were cancelled after at least one percent completion, and four additional units were deferred indefinitely (United States Department of Energy, 1988a: 21). Some utilities owned completed or nearly completed plants that might never open because of economic, political, managerial, and regulatory problems.

This book examines the development of the Shoreham nuclear power plant on New York State's Long Island, one of the most striking projects in the industry's history. The Long Island Lighting Company (LILCO) ordered Shoreham in 1967 in the early enthusiasm for nuclear power, but did not complete its construction until 1985. Shoreham still was not operating commercially in 1990, although costs for its construction, financing, and other expenses had exceeded $5.5 billion. In 1989 the United States Nuclear Regulatory Commission granted Shoreham a commercial operating license, but that year LILCO and New York State signed an agreement under which LILCO would sell Shoreham to the State for $1, which in turn would dismantle the plant. A fully functional electricity generating facility was never to be used, despite great uncertainties regarding future power supplies for Long Island.

In some respects the Shoreham case is unique in nuclear power's history. The duration of the controversy (at this writing almost twenty five years), the plant's eventual cost (At over $6,000 per kilowatt of capacity, the most expensive in the nation), and its impact on the politics and economics of a region of 2.7 million people were extraordinary. Shoreham also was a pivotal case testing the influence of Federal versus state and local governments in nuclear power policy, as state and local governments, attempting to "veto" the plant, stalemated Shoreham's Federal operating

Table 1. Nuclear Units Ordered and Cancelled Through 1982

Year	Orders Placed		Cancellations		Cumulative Orders*	
	Number	GWe**	Number	GWe**	Number	GWe**
Pre-1972	131	109.39	0	--	131	109.39
1972	38	41.31	7	6.12	162	144.59
1973	41	46.79	0	.00	203	191.38
1974	28	33.26	7	7.22	224	217.42
1975	4	4.15	13	14.70	215	206.88
1976	3	3.80	1	1.15	217	209.53
1977	4	5.04	10	10.81	211	203.76
1978	2	2.24	14	14.49	199	191.51
1979	0	0	8	9.55	191	181.96
1980	0	0	1	18.00	175	163.96
1981	0	0	6	5.78	169	158.18
1982	0	0	18	21.94	151	136.24
Total	251	245.99	100	109.75	151	136.24

Source: United States Department of Energy, *Nuclear Plant Cancellations: Causes, Costs, and Consequences* (April, 1983), Table 1.

*Cumulative outstanding orders
**Gigawatts-electric

license for several crucial years by refusing to cooperate in mandatory emergency planning. By 1989 many in the Federal government and the industry saw the prospective destruction of a new, completed, licensed plant to be a major blow to any attempt to revive nuclear power's development in the United States.

Despite its uniqueness, Shoreham's fundamental problems were shared with other nuclear projects in the 1970s and 1980s, and thus the case also exemplifies nuclear power's contemporary difficulties. During the period nuclear projects evolved among *technological*, *economic*, *managerial*, and *political* problems. None of these problems individually or collectively doomed Shoreham or any other plant. They constrained and influenced decisions, but ultimately actors made choices, for a variety of reasons, which tilted the histories of nuclear projects like Shoreham one way rather than the other. By examining Shoreham's history closely we can try to understand why some choices, and not others, were made; this will help us evaluate Shoreham's outcome. Similar dynamics, with different outcomes, may be found in other disputes over nuclear power or other complex technologies. The next section reviews the general technological, economic, managerial, and political contexts of these choices.

Technological Factors: Discoveries, Uncertainties, and Limits on Safety Strategies

Discoveries and Uncertainties

In the 1960s and early 1970s both the number of nuclear plants on order and their generating capacity increased greatly. It generally is recognized now that scientific knowledge and operating experience were not advanced enough to support the build-up. Cook (1985: 84) noted that

> [A]s the nuclear program gathered momentum, most of the new plants had to be scaled up--to maybe 1.2 million kilowatts, to retain their cost advantage [over coal]. In scaling up, however, the utilities were moving into unexplored territory. Nobody knew what the design problems were or, worse, what they were going to become. Says Christopher Young, senior vice president of Donaldson, Lufkin, and Jenrette, "The assumption was you had a mature technology when in fact it was still evolving" (1985: 84; see also Morone and Woodhouse, 1989).

Surprising discoveries and uncertainties appeared repeatedly during the expansion, increasing the plants' costs and diminishing public confidence in nuclear power. For example, concerns about the effectiveness of emergency core cooling systems jolted the industry in the early 1970s (Campbell, 1988: chapter 4). The Three Mile Island (TMI) accident in 1979 also took the industry and regulators by surprise, forcing a reassessment of construction, operating, and regulatory practices; numerous new regulations were applied both to existing and new plants (Komanoff, 1981).

Limits on Safety Strategies

The decision to build progressively larger nuclear reactors based on light water technology limited the safety strategies available to industry and regulators. When nuclear power became controversial in later years, defending these particular strategies was difficult because of the open-ended nature of safety objections which could be made against them.

Large nuclear reactors using light water technology came to dominate the industry here and in Europe. A number of alternative reactor types, some safer than others, were considered for development in the 1950s and 1960s. But light water technology had an "inside track" because private industry involvement was seen as essential to nuclear power's growth, and industry saw light water reactors as being closest to commercial development (Bupp and Derian, 1978). Some reactors which had major safety advantages faded, or were not pursued, because they could not compete economically with light water reactors. Morone and Woodhouse noted that "Technical, economic, and political considerations dominated early choices about civilian nuclear power. Reactors did have to pass a preliminary test of safety-in-principle, but beyond that were evaluated largely on the basis of economic suitability for a private nuclear industry. Safety was more a constraint than an objective in this process, and public acceptability was not directly addressed" (1989: 66-67).

Morone and Woodhouse viewed the dramatic increase in the number and generating capacity of reactors in the mid-1960s as fateful, because such reactors required a safety strategy which ultimately left the utilities in a very difficult political position. Possible safety strategies for nuclear power plants include *redundancies and error margins* to prevent errors from becoming emergencies; *systems to intervene* in the event of an emergency; *containment* to prevent radiation from reaching the wider environment; and *remote siting* to dilute radioactivity that might escape. The industry could rely confidently on containment and remote siting when dealing with few and relatively small reactors. Accidents in smaller reactors would not be powerful enough to destroy the containment system. Furthermore, reactors in the early 1960s commonly were not sited in populated areas because they were more demonstration projects than a major part of generating systems.

However, when the reactors' generating power increased greatly, containment and remote siting no longer sufficed as safety strategies. A worst-case accident in larger reactors might be powerful enough to breach containment. Also, while utilities ordered more nuclear plants because they were seen as being economically competitive with fossil-fuel plants, the plants were competitive partly because they were placed near populated areas. Utilities favor siting a plant near the population to be served to reduce transmission costs; remote siting would increase transmission costs, and might make a power plant inefficient economically. Remote siting, like containment, also might not prevent damages from a major accident at a relatively large and powerful nuclear plant (As the radiation from the Chernobyl accident demonstrates.) (Campbell, 1988: 52). Strategies to *prevent the possibility of accidents* thus had to figure more heavily than containment or remote siting in industry development.

But prevention strategies could never assure that all accidents *would* be prevented:

Now it became necessary to envision all the possible events and sequence of events, consequences, and consequences of consequences that could plausibly lead to meltdown...[t]he seemingly endless nature of the list was precisely the problem. The reliance on prevention created a regulatory, political, and analytical morass; it put nuclear decision makers in the position of having to demonstrate that they had anticipated everything that could go wrong in a reactor...[I]t was impossible to prove that the unanticipated would not occur (Morone and Woodhouse, 1989: 89, 91).

In summary, the rapid growth in the number and size of reactors in the mid-1960s (of which Shoreham was one) laid the groundwork for long-run problems. Knowledge about the operation of large nuclear plants was relatively weak prior to the industry build-up; the resulting surprises and uncertainties increased costs and diminished confidence in the industry. Also, when the atmosphere toward nuclear power grew hostile, it was difficult to argue in a conclusive way for the safety of large reactors using light water technology.

Economic Factors: Shifting Electricity Demand and Financing Problems

Shifting Electricity Demand

In the 1960s and early 1970s the demand for electricity was strong and rising, and so the need for the many nuclear plants on order appeared great. The first oil price increases early in the 1970s also helped ease for many people the doubts about nuclear power that were beginning to emerge, for nuclear power was seen as a potentially attractive alternative to imported oil. The Nixon administration's 1974 "Project Independence" called for nuclear power to provide 40 percent of the nation's power by 1990, and 50 percent by 2000 (Cook, 1985: 88).

But energy prices continued to rise substantially through the 1970s. Coupled with slower economic growth, this dampened the growth of electricity consumption (See Table 2). Many utilities and the State commissions regulating them now concluded that they probably would not need many of the facilities on order until far into the future, and nuclear projects were the most likely to be curtailed because of their higher economic and political risks.

Table 2. Percentage Changes in Electricity Generation and Gas/Electric Prices, 1960-1987

Year	Gas/Electric Price Change	Electricity Generation Change
1960	4.0%	6.5%
1961	.9	5.0
1962	.0	7.7
1963	.0	7.2
1964	.0	7.3
1965	.0	7.2
1966	.4	8.4
1967	.4	6.1
1968	.8	9.5
1969	1.7	8.5
1970	4.5	6.2
1971	6.7	5.3
1972	5.2	8.5
1973	4.9	6.3
1974	15.4	.3
1975	16.2	2.7
1976	11.5	6.3
1977	13.0	4.2
1978	8.9	3.9
1979	10.9	1.9
1980	17.0	1.7
1981	14.7	.4
1982	13.8	-2.3
1983	8.9	3.1
1984	3.8	4.6
1985	1.6	2.2
1986	-1.3	.7
1987	-1.8%	3.4%

Source: Prepared from U.S. price index for gas (piped) and electricity as reported in *Economic Report of the President*, 1989 (Table B-59), and from estimates of net generation of electricity by electric utilities, U.S. Department of Energy *Annual Energy Review*, 1987, Table 83.

Financing Problems

For a variety of reasons, nuclear projects frequently ran far beyond their expected budgets and timetables. Many utilities with nuclear projects experienced financially untenable situations because under regulatory accounting they could not recover costs until the delayed plants were actually operating.

Nuclear power is more complex than fossil fuel-fired technology, and the consequences of accidents potentially are greater; thus, its capital and labor costs were substantially higher. But uranium fuel costs less than oil or coal, and utilities argued that the fuel cost savings would more than offset the higher construction costs over the life of the plants. Overtime, however, the total costs of building nuclear plants skyrocketed. As the projects stretched out their *financing* problems became extraordinarily difficult, further crippling the industry.

Generally, state public utility commissions will not allow a utility to charge customers for the costs of a plant--even the interest costs of borrowed money--until the plant is actually in service. Once the plant is in service it is entered into the utility's "rate base," on which a public utility commission bases the amount of revenue that the utility can collect from customers. (The actual revenues collected reflect other factors as well, such as electricity demand).

Thus, in utility accounting the company imputes an interest charge to the money it has invested in a project based on its average cost of capital, adds that charge to the total investment, and reports it as a noncash credit in its income statement. The credit is called Allowance for Funds Used During Construction (AFUDC). Adding a 10-15 percent interest charge to a $2 billion dollar investment, for example, would increase the costs of the project by $200-$300 million in a given year. That interest charge, in turn, is added to the total investment, and itself earns a return in future years (Cook, 1985: 94).

AFUDC, while a commitment to future earnings if a plant operated, provides no current operating cash. As Cook notes, "As construction dragged on, AFUDC came to provide a major part of utility earnings--close to 100% for outfits like Lilco, 113% for Kansas Gas and Electric, and 126% for Public Service of New Hampshire. But, though the companies looked healthy, there was no cash behind these paper earnings." In 1985, current investment in ongoing nuclear projects accounted for between 49 and 78 percent of total net plant (the value of operating assets minus depreciation) for a number of utilities (1985: 94). In 1983, Shoreham's estimated costs were *205 percent* its total net plant (See Table 10, Chapter 4). Thus, a large share of utilities' investments were generating no operating cash to pay for ongoing project expenses for long periods.

To mitigate these cash flow problems many states began to allow utilities to recover a part of such investments by allowing some Construction Work in Progress (CWIP) into utilities' rate bases. But the amounts were small relative to the size of the investments. Furthermore, as project costs rose into the billions, as delays grew, and as nuclear plants became more controversial, it became increasingly untenable for public utility commissions to grant utilities' requests to add more CWIP to rate base. The cash flow problems, in turn, made it difficult for utilities to raise funds to complete projects: "Utilities in such a bind had to finance their construction programs out of the cash their operating properties were generating, and when that wasn't enough, borrow money, sell more common or preferred [stock] or even cancel the dividend [on common or preferred stock]. When inflation, high interest rates, and delay push the cost of a project into the billions, that often gets pretty difficult" (Cook, 1985: 94; also see Campbell, 1988: chapter 6; U.S. Department of Energy, 1983: 17-25).

Public utility commissions and utilities considered various ways to cope with these problems, short of cancelling the projects (Campbell, 1988: 104-106; U.S. Office of Technology Assessment, 1984: 52-57). But, given the technological dynamics around nuclear power, the slowing of electricity demand growth, and other factors discussed below, in many cases the projects ultimately were cancelled.

Managerial Factors

The technological and economic problems of nuclear power were considerable even under the best of circumstances. Many utilities and their contractors, however, worsened the problems dramatically by failing to manage plant construction effectively.

Building a nuclear powerplant was far more difficult than building a fossil fuel plant. In addition to requiring more material and almost twice as much labor, nuclear plants have many more components, such as valves and piping reinforcers. A fossil plant might have 4,000 valves, while a nuclear plant might have 40,000. A fossil fuel plant might have 5,000 pipe supports, while a nuclear plant--because of concerns about damage from earthquakes or other shocks to the system--might require tens of thousands of pipe hangers, supports, and restraints designed and installed according to exacting standards (U.S. Office of Technology Assessment, 1984: 119).

Also, the *quality* standards for nuclear construction are particularly high because unchecked system failures can produce catastrophes. Greater experience with operating plants and increasing concern about plant safety, particularly after the Three Mile Island accident, increased the number and rate of issuance of new Federal regulations in the late 1970s and 1980s. The regulatory changes meant that engineering and construction work frequently had to be modified or re-done (Komanoff, 1981; Office of Technology Assessment, 1984: 122-123).

Finally, a nuclear construction project required a larger number of organizations than fossil fuel projects, increasing the demands for coordination. Utilities generally felt that they did not have sufficient management experience to step into these projects unassisted, and thus hired other firms to design the plants and/or manage their construction. Other sub-contractors, in turn, would become involved, and eventually perhaps up to twelve contractors or sub-contractors might be working actively on site. Given the projects' technological complexity and duration, and the number of different actors involved, there was always a risk that different activities would not be linked effectively, controls would break down, work would have to be re-done, and delays and costs accordingly would increase. A major management challenge for the utilities and their construction managers was to keep these things from happening.

Some firms coped with these potential problems more effectively than others. Tables 3 through 6 summarize indicators of project success for the nuclear industry and selected nuclear plants. Table 3 presents the status in 1988 of nuclear projects arrayed by years in which the plants were ordered. Tables 4 through 6 present the construction and operating records of selected plants or the industry as a whole.

As Table 3 indicates, the vast majority of the nuclear plants ordered from 1953 through 1970 were operating by 1988, or retired after operation. (Three such early reactors were retired by 1988. Shoreham was one of the plants ordered in the 1960s but not operating by 1988). Twenty-nine reactors ordered from 1971 through 1973 were either operating, under construction, or had received an operating license, while seventy-one plants ordered in those years were cancelled or deferred indefinitely. By 1988 *all* of the plants ordered after 1973 were cancelled or deferred indefinitely. Thus, the industry was relatively successful in completing projects ordered from 1953 through 1970, suffered a downturn with plants ordered in the early 1970s, and had no success with plants ordered after 1973.

The difficulties of construction with *completed* plants varied greatly. Table 4 contrasts construction records for selected nuclear plants. The data need to be interpreted carefully because they reflect, in some cases, intentional stretching out of projects, some licensing delays unrelated to managerial performance, inflation, and accounting variations. Yet, the differences are so large that there is good reason to believe that they are real. Some examples, noted the Office of Technology Assessment, indicate that "Large and complex nuclear units can be built within budget and on schedule...These cases indicate that it is possible to construct and operate

Table 3. Status of Nuclear Units as of 1988 Listed by Year of Order (Number of Units in Different Status)

Year of Order	Operating or Retired after Operation	Cancelled[1] or Indef. Deferred	Shut Down	Under Construc.	Operating License Issued
1953	1				
1955	2				
1956	1				
1958	1				
1959	1				
1962	2				
1963	4	1			
1965	6		1		
1966	16		4		
1967	24	2	3	1	1[2]
1968	10	4	2		
1969	4	3			
1970	9	1		4	
1971	8	12		1	
1972	7	27		2	2
1973	7	32		1	1
1974		28			
1975		4			
1976		3			
1977		4			
1978		2[3]			

1. Includes plants rejected by New York State.
2. Entry is the Shoreham plant.
3. Two plants ordered but not in licensing process as of 1988.

SOURCE: Prepared from U.S. Department of Energy, Energy Information Administration, **Commercial Nuclear Power 1988** (September 21, 1988), Table E1.

Table 4. Construction Records of Selected U.S. Light Water Reactors

Plant (Year Ordered)	Construction time (yrs)	Year of comm. operation[1]	Cost[2]
Shortest construction times			
St. Lucie 2 (1972)	6	1983	$1,800
Hatch 2 (1970)	7	1979	607
Arkansas Nuclear 2 (1970)	7	1980	308
Perry 1 (1972)	8	1985	2,200
Palo Verde 1 (1973)	8	1984	1,900
Byron 1 (1971)	8	1984	1,500
Callaway 1 (1973)	8	1984	2,500
Longest construction times			
Diablo Canyon 1 (1966)	16	1984	$1,700
Diablo Canyon 2 (1968)	14	1984	1,700
Salem 2 (1967)	13	1981	704
Zimmer 1* (1969)	13	1985	2,400
Midland 1* (1968)	13	1985	2,700
Sequoyah 2* (1968)	12	1982	740
Watts Bar 1* (1970)	12	1984	1,500

1. Based on construction permit to commercial operation. Times are either actual or expected date of commercial operation.

2. Cost per kilowatt, actual or expected, in mixed current dollars. Because costs are mixed current dollars, cost figures are not comparable directly.

*According to a later report by the U.S. Department of Energy (1988: 87-92), the Zimmer 1 reactor was cancelled in 1984; the Midland 1 reactor was cancelled in 1986; the Sequoyah 2 reactor was shut down, with re-start scheduled for May 1988; and as of 1988 the Watts Bar 1 reactor was under construction.

SOURCE: U.S. Office of Technology Assessment, 1984: 115. Information on date of plant order from U.S. Department of Energy, *Commercial Nuclear Power 1988* (September 21, 1988), Table E1.

nuclear powerplants efficiently and reliably...even in the most difficulty regulatory environment." On the other hand, "other plants have had a long history of problems" (1984: 113, 115).

Similarly, nuclear plants' *operating performance* varied substantially. A power plant's "capacity factor" refers to the percentage of time that a plant has operated at its design potential over a specified period. No plant can operate at 100 percent capacity over any significant length of time because of the inherent need for maintenance and refueling. Since 1973, *overall* capacity factors in the nuclear sector generally have remained in the range of 55-63 percent (See Table 5). In comparison, the normal operating capacity for comparably-sized coal-fired plants is about 60-70 percent (Komanoff, 1981: 253). But some plants operated far more efficiently than others, some with lifetime capacity factors near eighty percent and others near 35 percent (See Table 6).

There are no obvious explanations as to why some utilities' nuclear plants were built and/or operated relatively efficiently, while other utilities' plants threatened to destroy their owners. However, some research suggests that utilities' performance with nuclear plants tends to improves with experience.

Charles Komanoff's multivariate study of all U.S. nuclear reactors completed from December 1971 through December 1978 identified patterns in construction costs (He studied a total of 46 reactors; four reactors sold for fixed prices were omitted from the research) (1981). Among other findings, Komanoff reported that a doubling of nuclear construction experience for a plant's Architect/Engineer lowered a plant's construction costs somewhere between five and nine percent. Also, the building of multiple units on a site, with related learning and other effects, reduced costs for each unit somewhere between two and seventeen percent.

Variables capturing broader political forces, however--such as the growing concern over nuclear plants' safety--had four or five times the statistical effects of these "experience-related" variables (See Komanoff's discussion of his Nuclear Sector Size and Cooling Tower variables [1981: 201-204]). One could conclude that political and technological forces pushed costs upwards, but that experienced firms were able to cope with these pressures somewhat better than others.

A study of forty-one utilities by Osborn and Jackson reinforces this impression. Osborn and Jackson found a strong association between the share of a utility's capital investment in nuclear power and the operating reliability of the utility's nuclear plants, although they also stressed that high operating reliability did not necessarily imply the safety of the plants (Plants may operate, for example, even though safety systems are weak.) (1988). Alfred Marcus, in a study of thirteen utilities, similarly found that

Table 5. Nuclear Power Plant Operations

Year	Operable Units	Nuclear Electricity Generation, Million Net Kilowatt Hours	Capacity[1] Factor
1973	39	83,479	53.7%
1974	48	113,976	47.9
1975	54	172,505	56.0
1976	61	191,104	54.9
1977	65	250,883	63.4
1978	70	276,403	64.7
1979	68	255,155	58.5
1980	70	251,116	56.4
1981	74	272,674	58.4
1982	77	282,773	56.7
1983	80	293,677	54.4
1984	86	327,634	56.3
1985	95	383,691	58.0
1986	100	414,038	56.9
1987	107	455,270	57.4
1988	108	526,901	63.5
1989[2]	110	391,614	61.8

1. Monthly capacity factors were computed as the actual monthly generation divided by the maximum possible generation for the month. The maximum possible generation is the number of hours in the month multiplied by the monthly net summer (peak) capability. This fraction is then multiplied by 100 to obtain a percentage. Annual capacity factors are averages of the monthly values for that year.

2. Figures for 1989 are nine-month totals. Nine-month capacity figures for 1987 and 1988 were 57.8% and 64.6%, respectively.

Source: Prepared from U.S. Department of Energy, *Monthly Energy Review*, September, 1989, Table 8-1.

Table 6. Performance of Selected U.S. Light Water Reactors

Plant (Year Ordered)	Lifetime capacity factor	Design capacity (MWe)	Years of Operation[1]
Best capacity factors[2]			
Point Beach 2 (1967)	79%	497	11
Connecticut Yankee (1962)	76	582	15
Kewaunee (1967)	76	535	9
Prairie Island 2 (1967)	76	530	8
Calvert Cliffs 2 (1967)	75	845	6
St. Lucie 1 (1967)	74	830	6
Prairie Island 1 (1967)	71	530	9
Monticello (1966)	71	545	12
Vermont Yankee (1966)	70	514	10
Calvert Cliffs 1 (1967)	70%	845	8
Worst capacity factors			
Beaver Valley 1 (1967)	34%	852	7
Palisades (1966)	39	805	11
Davis Besse 1 (1968)	40	906	5
Brunswick 2 (1968)	41	821	7
Salem 1 (1966)	46	1090	6
Indian Point 3 (1967)	46	965	6
Brunswick 1 (1968)	48	821	6
Rancho Seco (1967)	50	918	8
Duane Arnold (1968)	51%	538	8

1. Years of operation by January 1983.

2. Includes only plants greater than 100 MWe in operation three years or longer.

SOURCE: U.S. Office of Technology Assessment, 1984: 117. Information on date of plant order from U.S. Department of Energy, *Commercial Nuclear Power 1988* (September 21, 1988), Table E1.

the commitment and effectiveness of utilities' nuclear management varied considerably. Utilities with better records, in feedback fashion, tended to improve their performance further, while utilities with poorer records tended to deteriorate. However, he found no association between the *absolute* amount of a utility's nuclear generating capacity and the operating efficiency of it's nuclear plants (1988). Finally, after reviewing a series of studies on plant construction costs, Richard Lester suggested that there was moderately strong evidence that "the lack of experience in the nuclear power plant supply sector has affected its overall performance," and that "the projects that had apparently been successful were characterized by project teams with substantial prior nuclear construction experience, whose individual members had not overstepped their capabilities and had quickly recognized the need for a fundamentally different approach [to construction] in the case of nuclear plants" (1986: 371-372).

Even if utilities' with more experience in nuclear power coped with their problems somewhat more effectively than less experienced utilities, the problem was surviving long enough to get the experience. Companies beginning nuclear programs in the late 1960s-early 1970s would have had to manage their programs *well enough* to negotiate a relatively unforgiving environment in the late 1970s and 1980s. The results depended on their managerial skills and the level of environmental difficulties. Some companies succeeded, and others did not.

Political Attitudes

Elections, public and private leaders' personal commitments and ambitions, questions of intergovernmental relations, and many other factors influenced the politics of nuclear projects. Attitudes of key groups and the public toward nuclear power heavily shaped these other political factors; political events, in feedback fashion, then further modified attitudes.

Political attitudes helped determine how much credibility utilities would have in coming to grips with nuclear power's special technological, economic, and managerial problems. One view of nuclear power saw it as an evolving but valuable resource; surprisingly expensive but competitive technologically with fossil fuels; and nuclear management, while flawed, improving with experience. Another view saw nuclear power as inevitably and unacceptably risky; its costs uncontrollable; and its management as inherently flawed because of the technology's special dangers and, in many cases, corporate irresponsibility (Perrow, 1984). Politically active groups backed each of these positions. Winning the support of key governmental leaders, and shaping the direction of voters' opinions, influenced governmental policy toward a given nuclear project.

Two important characteristics of public opinion on nuclear power are that (1) scientific opinion tends to support the development of nuclear power, while broader public opinion tends to oppose it; and (2) the level of public support for nuclear power declined sharply from the 1970s to the 1980s.

Differences in Scientific and Broader Public Attitudes

Much of the public came to define nuclear power as something to be especially feared and unacceptable. This constrained nuclear power's expansion. Yet, at the same time, scientific opinion generally supported nuclear power's development, keeping arguments in favor of the technology intellectually respectable and helping to keep the industry alive.

For most people, perceptions of risks reflect the characteristics of hazards more than objective measures of mortality or morbidity. Research suggests that people's perceptions of risk vary systematically across two factors--the extent to which the risk is "dread" and the extent to which the risk is well-known or understood. The components of these factors are summarized below. Slovic, Fischoff, and Lichtenstein (1986) note that the extent to which a risk is "dread" affects attitudes particularly strongly; the higher a risk is on the "dread" factor, the more people want to see its current risks reduced through tighter regulation or other means.

Factor 1: Dread Risk

Controllable	Uncontrollable
Not dread	Dread
Not global catastrophic	Global catastrophic
Consequences not fatal	Consequences fatal
Equitable	Not equitable
Individual	Catastrophic
Low risk to future generations	High risk to future generations
Easily reduced	Not easily reduced
Risk decreasing	Risk increasing
Voluntary	Involuntary
Doesn't affect me	Affects me

Factor 2: Unknown Risk

Not observerable .... Observable
Unknown to those exposed...Known to those exposed
Effect delayed .... Effect immediate
New risk... Old risk
Risks unknown to science .... Risk known to science

Source: Slovic, Fischoff, and Lichtenstein (1986: 9).

Nuclear power is perceived to rank highly on the components of both "dread risk" and "unknown risk." For example, it is considered to present high catastrophic potential and high risk to future generations, to be involuntarily assumed, to have inobservable and delayed effects, and to present risks which generally are not well-known to science.

It is also recognized that experts focus more directly on expected annual mortality when judging riskiness and the need to regulate risks. One of the most striking features of the politics of nuclear power is that most scientists, particularly those working in the energy field, and largely regardless of economic interests or political ideology, reject the view that nuclear power is an unacceptably risky technology. One study asked four groups--members of the League of Women Voters, college students, Active Club (a business and professional group) members, and experts--to rank 30 activities and technologies according to their level of riskiness. The League of Women Voters and college student groups ranked nuclear power as the most risky of the thirty activities and technologies, and Active Club members ranked it eighth in risk. In contrast, the experts ranked it twentieth out of the thirty (Slovic, Fischoff, and Lichtenstein, 1986: 6).

Similarly, Stanley Rothman and S. Robert Lichter surveyed 1,136 members of "leadership groups" in 1982, including national-media journalists, science journalists, military leaders, congressional staff, partners in top corporate-law firms, creators of television and motion picture entertainment, high-level government bureaucrats, and public interest-group leaders; and 925 scientists drawn from *American Men and Women of Science* in 1980. They examined how views of the safety of nuclear power plants varied across the groups (See Table 7). A minority of the total Leadership sample considered nuclear power to be relatively safe. In contrast, a majority of the scientists, and a large majority of energy experts, considered nuclear power to be relatively safe. Ninety percent of the

Table 7. Percentage of Respondents Rating Nuclear Plants as Safe*

Sample Groups	Percentage Rating Nuclear Plant Safety 5 or Higher*	Sample Size
Total Leadership sample	36.8%	1,203
Bureaucrats	52.0	199
Congressional aides	39.1	132
Lawyers	48.6	149
Media	36.5	156
Journalists at *New York Times* and *Washington Post*	29.4	51
Journalists at TV networks	30.6	49
Military	86.0	152
Movies	14.3	90
Public interest	6.4	154
TV, Hollywood	12.5	103
Total Scientists sample	60.2%	925
Energy experts	75.8	279
Nuclear energy experts	98.7	72

* Responses of 5 or higher on a 1-7 scale were considered as indicating relative confidence in nuclear plant safety.

Source: Rothman, S., and S. Robert Lichter, "Elite Ideology and Risk Perception in Nuclear Energy Policy," *American Political Science Review*, 81: 383-404 (1987).

general sample of scientists, and ninety-five percent of the energy experts, also believed that nuclear development should continue (p. 386).

Arguably, energy experts might favor nuclear power because of their employment or other economic interests, or political ideology. Rothman and Lichter examined the extent to which energy experts' employment status (business, government, pure academic, or consulting academic), region of residence, religous background, and self-described political beliefs accounted for variance in their opinions on nuclear safety. These variables accounted overall for only 11.6 percent of the variance in energy experts' safety estimates. Experts' estimates did not vary by their employment type, and political ideology explained only 5.3 percent of the variance in the estimates. In contrast, within the total Leadership sample, political ideology *itself* explained 41.2 percent of the variance in estimates of nuclear safety. "In short," they concluded, "scientists are mostly sanguine about nuclear safety, and economic self-interest plays no measurable role in their assessments. Ideology is a statistically significant predictor but the proportion of variance it explains is...very small as compared to that of nonscience leadership groups" (1987: 386-387).

Yet, the broader public perceives that scientists differ sharply regarding nuclear power's safety (Carl, 1972). Rothman and Lichter noted a poll of the general population in 1984 indicating that "almost 6 out of 10 respondents expressed the belief that those scientists who are energy experts are evenly split as to the safety of nuclear energy or consider nuclear energy unsafe." They maintained that the impression of division in the scientific community is created partly because scientists opposing nuclear power use popular media more extensively than other scientists, and partly because journalists seek out anti-nuclear scientists more actively than they seek out other scientists. "Indeed," they wrote, "the elite journalists we surveyed were asked to list information sources they considered reliable on nuclear energy. They selected more antinuclear than pronuclear sources by a margin of over three to two...the U.S. public has probably been influenced in its perceptions by the manner in which the risks of new technologies have been dealt with in the media" (pp. 390, 394, 395; see also Gamson and Modigliani, 1989: 18, fn. 16).

The Decline in Public Support for Nuclear Power

Public support for nuclear power declined through the 1970s, and sharply after the Three Mile Island accident in 1979.

During the 1970s certain events and issues shook nuclear power. In particular, several scientists from the Atomic Energy Commission criticized the agency's reactor safety program early in the 1970s, leading to hearings on emergency core cooling systems which left doubt in many peoples' minds about reactor safety (Campbell, 1988: chapter 4). The media's coverage of nuclear power increased in both quantity and critical tone about the time that public support for nuclear power began declining (Table 8). A circular process probably operated. The growing public suspicion of nuclear power reinforced production of stories further delegitimizing the industry, leading to further declines in public support. Gamson and Modigliani, in their study of media discourse and nuclear power, made a similar point. They wrote that "By examining discourse and public opinion as parallel systems, we deliberately avoid making certain causal assumptions. We do not, in this paper, argue that changes in media discourse *cause* changes in public opinion. Each system interacts with the other: media discourse is part of the process by which individuals construct meaning, and public opinion is part of the process by which journalists and other cultural entrepreneurs develop and crystallize meaning in public discourse" (1989: 2).

After the Three Mile Island accident in 1979 the percentage of the public opposing the construction of new nuclear power plants exceeded the percentage favoring it for the first time (about 40 percent favoring it, and 44 percent opposing it). For the next two years the percentage favoring exceeded the percentage opposing, but by 1982 opposition again exceeded support, and by about a two to one margin (U.S. Office of Technology Assessment, 1984: 24). While TMI was a major event, public support for nuclear power already had been declining for several years. Gamson and Modigliani suggested that "When public opinion is viewed over a 15-year period beginning in the early 1970s, TMI looks like little more than a small blip, which slightly accelerated a secular trend against nuclear power" (1989: 31).

Summary and Outline of the Book

All nuclear projects faced the technological, economic, managerial, and political problems discussed above. As the data in Tables 3 through 6 indicate, however, the projects' fates varied greatly. State and local politics, corporate histories, individual personalities, and even simple chance heavily influenced how general problems played out in specific cases. This book looks at how this process shaped Shoreham's history. One obvious lesson is that there *is* a politics of nuclear power (Wilson, 1980: 357). There was nothing inevitable on technological, economic, or managerial grounds about Shoreham not operating successfully. The decision to abandon the plant was a political decision, constrained but not determined by these other factors. Individuals will differ on whether the political decision was sound public policy.

Four primary sources of information are used in this book. First, the published literature on nuclear power sets the general industrial and

Table 8. TV News Coverage and Public Opinion on Nuclear Safety

Time[1] Period	Stories[2] Per Year	Reactor Safety[3] Stories (%)	TV[4] Slant	Public[5] Opinion %
1971				32% (57-25)
1970-75	26	8	-14 (25-39)	
1975]				23 (55-32)
1976-78	50	17	-24 (12-36)	
1978				2 (43-41)
1979	518	41	-30 (17-47)	
1980				-35% (28-63)

1. Some of the data are time-discrete and some are time-aggregate. Figures were unavailable for 1969.

2. Average yearly number of nuclear energy stories.

3. Percentage of all nuclear-related stories dealing with safety of nuclear reactor operation.

4. Percentage of positive stories minus percentage of negative stories.

5. Percentage in favor minus percentage opposed, as reported in national Harris and Gallup surveys.

Source: Reproduced from Rothman and Lichter, 1988: 948

political context in which Shoreham evolved. Second, I have used the large amount of archival information which has accumulated over Shoreham's life. Because the project was so controversial its evolution was recorded in detail in the operating archives of government agencies, the courts, legislatures, LILCO, and other organizations. This information includes reports and statistics, memoranda, public statements, and judicial and administrative decisions. A particularly valuable source of information was the archives of the five-year investigation by the New York State Public Service Commission (PSC) of the Shoreham project, and also the PSC's review of LILCO's proposed Jamesport nuclear project. Additional information was added to the narrative as it became available.

Third, the New York press, particularly the principal Long Island newspaper *Newsday*, followed Shoreham closely. *Newsday*, and to a lesser extent the *New York Times*, were excellent daily logs of Shoreham's chronology. Actors used these newspapers as conduits for information and views they wanted publicized, but the press functioned as much more than a messenger. Shoreham was the single most important issue on Long Island in the 1980s, and *Newsday* chronicled its complicated history very closely. Fourth and finally, I interviewed a number of individuals who had experience with the history of the plant.

The book is outlined as follows. Chapter 2 reviews the development of nuclear power leading up to LILCO's announcement of Shoreham, and discusses formative issues in the plant's design and overall project organization. During the period the industry and the government were both optimistic about the future of nuclear power, although in the late 1960s--just as key decisions about Shoreham were being made--signs of trouble emerged. Chapter 3 examines the early politics of Shoreham. LILCO's application for a construction permit for Shoreham resulted in the longest and most contentious permit hearings in the Atomic Energy Commission's history. A small but intense anti-nuclear movement on Long Island was stimulated by, and effectively used, these hearings. The hearings also upset the project's early timetable and put the Atomic Energy Commission and later the Nuclear Regulatory Commission on notice that Shoreham had to be held to a very high standard of regulatory review.

This demanding regulatory, political and legal environment required a high level of management skill during construction, a level, Chapter 4 suggests, the project never approached. LILCO's problems were so severe that the New York State Public Service Commission refused to allow LILCO to recover $1.4 billion of Shoreham's costs through electric rates on the grounds that the costs were incurred imprudently. Mismanagement of construction and regulatory relations, grotesque cost overruns, and a few conspicuous failures (such as the breaking during test runs of the plant's emergency power system) damaged LILCO's credibility as an operator of a nuclear plant.

Arguably, by the time Shoreham was completed the Long Island Lighting Company had developed the technical capabilities and experience to operate the plant effectively, and Federal regulators had scrutinized the plant more closely than any other in the nation. However, by that time the company was deeply vulnerable politically and financially, and leaders of State and local governments had vowed to block Shoreham's opening. Chapter 5 discusses how New York State and Suffolk County attempted to "veto" Shoreham's operating license by refusing to participate in Federally-mandated emergency planning exercises, and how the Federal government responded.

By 1988 LILCO was searching for a way out of the Shoreham controversy that would leave it as an investor-owned and solvent utility. Chapters 6 and 7 are concerned with the late negotiations and manuevering involving LILCO, New York State, Suffolk County, the Federal government, and other actors. During 1988 and 1989 LILCO moved close both to opening the plant and to bankruptcy; on several occasions the controversy seemed to close only to reopen; and the political commitments to stop Shoreham dominated the underlying technological and economic issues.

As discussed in this Introduction, probably all nuclear projects had difficult technological, economic, managerial, and political problems; some survived, others did not. Chapter 8 summarizes the factors that turned Shoreham into one of the industry's most spectacular and important failures, and what this history suggests about the dynamics of such complex and expensive disputes.

Chapter 2

SHOREHAM'S BEGINNINGS

The Early Development of Nuclear Power

About forty years ago the Federal government undertook a long-term effort to develop nuclear power, and eventually to draw private industry into the field. By the 1960s it appeared that nuclear power was likely to be the power source of the near future. Orders for new plants dramatically increased by 1968; Shoreham was one of these.

Early Federal Support for Nuclear Power

After World War II nuclear technology was regarded as promising, but still very risky economically to American industry. Consequently, the Federal government took the lead developmental role, trying to draw industry into cooperative projects to develop reactors and related technology. These early projects were extensions of military programs, but over time were expanded to include solely civilian nuclear projects.

The Atomic Energy Act of 1946 called in part for the peaceful application of nuclear energy, establishing the Atomic Energy Commission (AEC) to oversee its development. The early AEC's research, development, and production activities, however, were oriented mainly to military applications of nuclear power. Most of the funds spent on atomic energy went to military programs. Their military nature meant that the projects had to meet strict secrecy standards, limiting the extent to which industry could or would become involved.

David Lilienthal, the AEC's first Chairman and former director of the Tennessee Valley Authority, maintained that industry ought to have more access to information about atomic energy. In 1947 the AEC established an Advisory Group to suggest ways to involve industry in reactor development, and in 1948 the Group successfully argued that the Atomic Energy Commission should undertake a formal Reactor Development Program, with the Federal government financing industry's participation. Up to this time there were six American reactors operating--two in Chicago (built in 1942 and 1944); at Oak Ridge (1943); at Hanford, Washington (1943-44); and two at Los Alamos Laboratories in New Mexico (built in 1944 and 1946). The primary purpose of these reactors was research and development. Under the new Reactor Development Program the AEC brought in the Westinghouse Electric Company to help Argonne National Laboratories design and construct a reactor usable in a submarine. The first atomic-powered submarine--the Nautilus--was launched in 1954.

The Reactor Development Program supported other experiments in the commercial production of electricity involving Westinghouse, General Electric, and North American Aviation. Still, the AEC remained the main actor. The AEC conducted most of the basic research and development work at government-owned laboratories and test sites. While it financed industrial participation, it directed the projects and controlled the resulting information.

In 1950 the President of Monsanto Chemical Company proposed that "American Industry design, construct, and operate one or more atomic power plants with its own capital," with the government providing uranium to fuel the plants. Such a facility would generate electricity commercially, and also provide plutonium to the government for defense and research. The Atomic Energy Commission allowed Monsanto to study the AEC's reactor projects before proceeding to develop its own reactor. This was the first occasion in which the Commission allowed access to classified documents on atomic energy.

The AEC subsequently issued a "general invitation" to industry to study "dual purpose" reactors which would both generate electricity and produce plutonium for weapons. Detroit Edison and Dow Chemical, Monsanto Chemical and Union Electric, Pacific Gas and Electric, Bechtel Corporation, and other organizations proposed to build such reactors. In 1952 the AEC issued a second invitation for industrial participation, and soon the number of firms involved increased to twenty five.

In 1953 the Joint Committee on Atomic Energy (JCAE) held hearings on the Program. The Committee concluded that industry ought to participate more heavily, but needed more incentives than presently available. For example, at the time only the AEC could own nuclear facilities, and the JCAE suggested that the 1946 Atomic Energy Act be amended to allow utilities or other firms to own such facilities. The Committee also argued that the government would have to support related research more generously.

The Atomic Energy Commission then announced that it would begin supporting the development of facilities solely for the production of electric power, even if they had no military purpose. In 1954 the Atomic Energy Act also was amended to permit private ownership of reactors under AEC license, and to allow freer distribution of information on atomic research. The AEC would retain ownership of nuclear fuels, but would license the fuels to industry. In 1955 the AEC established the Power Demonstration Reactor Program. Under this Program the AEC would assist industry technically and financially, but the companies would actually build, own, and operate the plants. Industrial groups submitted proposals for plants in northwestern Massachusetts (The Yankee Atomic Power Plant), Nebraska, and Michigan. The Yankee plant was scheduled to begin operation in 1957 and cost $35 million. It began full-power operation in 1961, at a cost of $55 million. The Nebraska reactor went into operation in 1963, with a final cost of $40 million compared to an estimated $24 million. The Michigan reactor went into operation in 1964, at the originally estimated cost of $43 million.

Other companies applied for licenses to construct nuclear power plants at their own expense, without AEC assistance (In Dresden, Illinois; Indian Point, New York; and Vallecitos, California). The AEC approved all of the applications. The AEC also offered to finance industry construction of several reactors to be owned by the government, with the government to give utilities the option to purchase the steam produced by the reactors.

In 1956 several members of the Joint Committee on Atomic Energy maintained that still more was needed to accelerate commercial development of nuclear power. While legislation designed to do so failed in the House of Representatives (after passing the Senate), the AEC subsequently increased its own efforts, calling for construction of additional reactors. For example, it offered to finance all research and development costs and waive uranium lease charges for periods of up to seven years (Del Sesto, 1979: 54-56; Mazuzan and Walker, 1984: 77-83).

Developmental Problems

Liabilities From Nuclear Accidents

In 1957 civilian nuclear power's risks were publicly, conspicuously noted for the first time. Through the AEC, Brookhaven National Laboratories released *Theoretical Possibilities and Consequences of Major Accidents in Large Nuclear Power Plants*. The study indicated that a major accident could cause thousands of deaths and injuries at up to seventy-five miles from the accident, and that property damages could go as high as $7 billion (U.S. Atomic Energy Commission, 1957). Industry conceivably would not invest in nuclear power in light of such potentially devastating liabilities. The liabilities were so large that utilities and vendors might not even be able to obtain insurance for nuclear power plants.

Congress tried to ease these concerns in 1957 by passing the Price-Anderson Act, amending the Atomic Energy Act. The Price-Anderson Act required that licensees of nuclear facilities carry the maximum liability insurance coverage. However, it also limited claims against those held liable for a nuclear accident to $560 million, and provided up to $560 million in government indemnity to those held liable. In effect, the Federal government underwrote the complete liability of nuclear plant licensees or vendors. The Price-Anderson thus mitigated the threat of major liabilities from nuclear accidents, and lowered a barrier to private investment.

Still, by the early 1960s other concerns about the acceptability of nuclear power had developed. The Brookhaven study in 1957 noted the possibly devastating effects of a nuclear accident. Organized labor had questioned the safety of reactors for both workers and the surrounding communities. Safety concerns regarding the siting of reactors in populated areas emerged. Many groups began to question the adequacy of the AEC's standards for protection against radiation (Del Sesto, 1979: 66; Mazuzan and Walker, 1984: ch. III-IV, VII-VIII, XI).

Economic Problems

In the early 1960s the economics of nuclear power were still unfavorable, and possibly highly unfavorable. Nuclear projects ran past timetables and exceeded budgets. While it was technologically feasible to produce electric power from nuclear energy, the costs of doing so were not competitive with the costs of producing electricity from coal, hydropower, or other means. Del Sesto notes that up to this time "nearly every plant had cost more than originally intended, with many nearly doubling in price by the time they went on line. Even the lowest generating costs achieved were reported to be some fifty-percent higher than seven mills per kilowatt hour, the point at which nuclear facilities were seen to be competitive with conventional plants." Also, the increasing flow of cheap Middle Eastern oil, the declining price of coal, and improved plant efficiency had offset rising labor and materials costs for other types of power, weakening nuclear power's competitive position.

Another problem was that utilities at that time were far more likely to compare the costs of using different suppliers of particular types of fuels (coal, oil, etc.) rather than make more fundamental comparisons among alternative electrical technologies (e.g., nuclear versus fossil fuel); "therefore, the commercialization of nuclear energy faced the formidable problem of competing with mature technologies that had been tried and tested by utilities, a problem of course worsened by the fact that performance, particularly economic performance, had to rely primarily on predictions and extrapolations rather than on firm operating experience." Finally, the utilities seemed primarily interested in building plants with far greater generating capacities than those demonstrated feasible for nuclear technology (Del Sesto, 1979: 78-79). Thus, by the early 1960s, industry still had to be convinced of the reasonableness of undertaking multimillion dollar commitments to nuclear energy that might offer only limited savings --if any--over energy generated from fossil fuels.

The Surge in Orders

In 1962 the Atomic Energy Commission announced a major new effort to give industry incentives to build competitively large nuclear plants. The AEC would pay for research and development for plant design, although it would pay only ten percent of the pre-construction research and development costs of the plants. The utilities also would have to provide the sites on which the plants would be built, pay for construction, and be prepared to operate the facilities for at least five years after start-up. The AEC received proposals for construction of nuclear plants at Haddam, Connecticut (the Connecticut Yankee plant), and at San Onofre and Malibu, California. The Malibu plant was never completed because of technical problems and public opposition to siting a plant near an earthquake fault. The Connecticut Yankee and San Onofre plants were completed. Also, in 1963 and 1964, the Congress amended the Atomic Energy Act to allow industry to invest in and own nuclear fuels for the first time (Del Sesto, 1979: 81-85).

By 1962 the AEC was forecasting that nuclear power would soon be economically competitive, and an event in late 1963 appeared to justify the AEC's optimism. The Jersey Central Power and Light Company announced that it would purchase a 515 MW nuclear plant from General Electric at a fixed price of $66 million. The plant would be sited at Oyster Creek, New Jersey. For the $66 million, plus certain inflation-related adjustments, General Electric would build the facility and obtain the necessary construction and operating permits. Once the plant was completed the utility would simply have to "put the key in the door and crank up the generating equipment;" thus, such plants came to be known as "turnkey" plants.

Jersey Central maintained that in five years the Oyster Creek plant would generate electric power at a price more favorable than a plant fired by any other type of fuel, including coal. It said that for a coal-fired plant to be competitive at the Oyster Creek site, the cost of coal delivered to the plant would have to be less than 20 cents per million British thermal units ($.20/mbtu). The cost of coal used by utilities at this ranged from $.35/mbtu in New England to less than $.20/mbtu near Appalachian coal fields. Jersey Central's costs were currently around $.29/mbtu.

The announcement suggested that investment in nuclear power had become competitive, and sometimes even superior, compared to investments in other power sources. In 1962 the Atomic Energy Commission had forecast that nuclear reactors beginning operation in 1980 would produce electricity for 3.8 mills/kwh during the first five years of their operation. The Oyster Creek plant, scheduled for operation in 1968, presumably would meet that goal twelve years earlier than 1980 (Bupp and Derian, 1978: 45).

Between 1964 and 1967 General Electric and then Westinghouse offered other turnkey plants for sale. By the time the companies stopped making the turnkey offers, General Electric had build seven turnkey plants, while Westinghouse received contracts for six more.

Arguably, vendors of nuclear plants believed that nuclear plants indeed could be built at the turnkey prices. First, the vendors, cooperating with the utilities, could draw on what was by that time a substantial amount of nuclear construction experience. Much of this experience was with facilities that were much smaller than the turnkey plants. However, vendors expected that they could transfer the knowledge acquired in building smaller facilities to larger, presumably more efficient facilities (Bupp and Derian, 1978: 185-186). Second, much of the equipment in a nuclear power plant--with exceptions such as the nuclear-generated steam supply system--resembled the equipment in a conventional coal or oil-fired plant. Although there were important differences between nuclear and fossil-fuel fired plants, vendors felt familiar with some of the technology. In later years it became evident that experience with smaller reactors could not be transferred efficiently to the construction of larger reactors, and that the qualitative differences between the construction of nuclear and fossil fueled-fired plants were larger than originally expected. However, at the time the assumptions were not generally questioned.

Jersey Central's favorable evaluation of Oyster Creek also made some optimistic assumptions regarding the plant's operations. Jersey Central assumed that the plant would operate at 640 MW, more than twenty-five percent higher than the plant's nominal rating of 515 MW. It also assumed that the plant could operate at eighty-eight percent capacity (The capacity factor refers to the amount of electricity produced relative to the amount that could be produced if the plant were operated non-stop throughout the year. No plant can operate at 100 percent capacity because of the inherent need for maintenance and refueling). The normal operating capacity for a coal-fired plant of comparable size was 60-70 percent (Komanoff, 1981: 253).

One of the few individuals criticizing the favorable analyses of nuclear power economics was Philip Sporn, the President of American Electric Power Company, the nation's largest coal-burning utility. Sporn maintained that the analyses favorable to nuclear power downplayed both the progress in fossil fuel-fired technology and the growing availability of fossil fuels.

Properly considering these factors would increase the attractiveness of conventional power sources, and diminish the comparative attractiveness of nuclear power. Also, he argued that Westinghouse and General Electric appeared to be taking substantial risks and/or very narrow profits on the early turnkey plants, risking "somewhat greater uncertainty in the costs behind the turnkey price than might be tolerable repeatedly...In addition, I am extremely doubtful, even in the case of the most fortuitous minimization of the effects of the manufacturer's financial risk, whether the Oyster Creek sale will yield the same financial return as the manufacturer would normally require as a matter of sound business; the kind of return he requires of his conventional turbine business, for example" (cited in Del Sesto, 1979: 88-89).

Yet, even Sporn's estimates of the costs of nuclear power did not differ greatly from those estimates made by its proponents. Sporn's estimates used data provided by the reactor manufacturers, and so the estimates naturally would converge somewhat. He also generally accepted the assumptions that it was possible to predict the costs of nuclear power plants which were much larger than the plants currently operating, and that the relative costs of plants would decline as manufacturers gained experience building them. Sporn's estimates of the capital costs of reactors differed from those of the manufacturers and the AEC by only about ten to fifteen percent. Bupp and Derian note that "...by early 1963, there was a consensus on the estimated costs of electricity from light water reactors. The government and the reactor manufacturers were at one extreme, predicting about 5.8 mills/kwh; and Sporn was at the other, predicting about 7.6 mills/kwh--a relatively small difference of about 25 percent. Contemporary costs of electricity from fossil fuels ranged between 4.3 mills/kwh and 6.4 mills/kwh. Hence, depending upon the assumptions one chose to work with, nuclear power was either 'quite competitive,' 'almost competitive,' or 'not yet competitive' in 1963" (1978: 47-48).

Eventually, turnkey plants did cost more to build than the stipulated price, costing about the same amount as non-turnkey plants constructed in the period, with the actual cost of building the plants double the initial price. However, "at the same time, they served their purpose by announcing, however prematurely, that nuclear power was competitive with conventional power generating technologies. The turnkey orders piqued the utilities interest and doubtless played a significant role in the great order surge for nuclear power that would occur in 1966-67, less than three years after Jersey Central's historical Oyster Creek contract with General Electric" (Del Sesto, p. 90; Bupp and Derian, p. 79).

Beginning in 1965 utilities began placing a large number of orders for nuclear plants without firm price guarantees, and Babcock and Wilcox and Combustion Engineering joined General Electric and Westinghouse as reactor manufacturers. By the end of 1967, U.S. utilities had ordered 71 nuclear plants with a total design capacity of 51,946 MW. In 1966-67 alone utilities ordered 51 plants, with a total design capacity of 42,984 MW; thus, 72 percent of the plants and 83 percent of the capacity to that point were ordered in those two years (U.S. Department of Energy, 1988: Table E1). The orders for nuclear plants in 1966-67 represented nearly half of all the new power generating capacity ordered by United States utilities in those years (Bupp and Derian, p. 49).

Philip Sporn called the major surge in orders for nuclear plants "the great bandwagon market." Four factors contributed to the surge in orders.

First, the coal industry saw little need to vigorously contest nuclear power's expansion. Demand for coal had increased in the early 1960s, and by 1965 coal prices were strong. The coal industry was already profitable; it had little immediate need to try to draw utilities away from nuclear power by offering very favorable prices to utilities contemplating a fuel shift. Also, the utilities themselves might have had some interest in diversifying their sources of power, to diminish their reliance on coal.

Second, concern about environmental pollution increased the demand for nuclear power. Local laws had begun requiring tighter controls on sulfur dioxide and other emissions from plants using fossil fuels. Low sulfur coal and oil were not readily available, and were expensive. Scrubbing systems to remove pollutants from stacks were expensive. People opposed the construction of long power transmission lines for a variety of reasons. Coal mining, particularly strip mining, generated its own environmental problems. The utilities believed that environmental regulations might come to favor nuclear power, and to discourage the use of fossil-fuel generated power.

Third, the Northeast blackout in 1965 led the Federal government to call for the interconnection and pooling of regional power systems. Regional pooling meant that, should a nuclear plant (or other type of plant) be shut down for one reason or another, a utility could call on pool members for additional power. Pooled financing meant that the risks associated with the construction of large nuclear power plants could be spread. Regional pooling correlated strongly with the move to nuclear power that occurred in 1966-1967 (Del Sesto, 1979: 91-92).

Fourth, the utilities felt that if they failed to move into nuclear power the AEC and the Joint Committee on Atomic Energy--both strongly interested in the development of nuclear energy--would foster the development of governmentally-owned power arrangements such as the TVA. Governmental control of large scale nuclear power systems might preempt the utilities' move into what appeared, to many, to be the power source of the future (Bupp and Derien, 1978: 74-75; Clark, 1985).

It was during the order surge of the mid-1960s that the Long Island Lighting Company announced the decision to build the nuclear plant at Shoreham.

The Decision To Build Shoreham

The Long Island Lighting Company (LILCO) was founded in 1910, and by 1930 was the major power company on Long Island. From 1930 to 1980 the population of Nassau and Suffolk counties increased from 460,000 to 2.7 million; over two-thirds of that increase occurred between 1950 and 1970. Both the spread and the increase of population increased the demand for electricity. LILCO and other utilities also promoted the consumption of electricity, encouraging construction of "all-electric" homes and, until 1956, even selling appliances. Also, LILCO had a firm lock on energy supply in general on Long Island because the company was the sole supplier of natural gas within most of the company's electric franchise area.

The increasing consumption of electricity put the utilities in an advantageous cycle. Per unit costs of production declined as the production of electricity increased. Declining per unit costs lowered the unit price of electricity, which in turn led to further increases in the demand for power. Between 1950 and 1960, Long Island's population doubled from 950,000 to 2 million, but LILCO's electricity sales more than tripled, from 1.2 billion to 4.4 billion kilowatt hours. Between 1960 and 1970 its sales almost tripled again, to 13 billion kilowatt hours, even though the number of customers increased by only one-third. Generating costs per kilowatt hour declined, from 4.9 cents in 1940 to 2.4 cents in 1970 (Diamond, 1981: 11R).

Like other utilities, LILCO was attracted to nuclear power in the mid-1960s. Since 1955 LILCO had belonged to the Atomic Power Development Associates, an organization conducting research and development on nuclear power plants, and since 1960 belonged to the Empire State Atomic Development Associates, funding related research and development. In 1965 LILCO established a Nuclear Engineering Division; the Division's staff was involved in seminars and classroom studies of nuclear power. Arthur D. Little, Inc. pointed out that "By 1967, LILCO had spent approximately $5.8 million (roughly equivalent to $18 million in 1984 dollars) in the above activities and was probably as well qualified as most utilities to embark on a program of nuclear power" (1984: 22). The surge in consumption allowed the utilities to invest heavily in new capacity, and LILCO wanted a large share of its new investment to be in nuclear facilities.

In 1966 LILCO announced plans to construct the nuclear plant at Shoreham. In 1967, it announced plans for another plant at Lloyd Harbor (plans cancelled in 1969) (*New York Times*, April 14, 1966: 22; October 10,

1967: 31). These announcements occurred in the excitement over nuclear power generated by the Jersey Central Power and Light Company's Oyster Creek plant and other turnkey offers. For example, in 1963 New York's Niagara Mohawk Power Company had announced a contract for the Nine Mile 1 plant, off Lake Ontario, for a turnkey price of $154 per kilowatt of capacity. Wilfred O. Uhl, LILCO's planning manager in the early 1960s, said that the Oyster Creek and Nine Mile 1 announcements "began to convince us of (atomic power's) economic feasibility" (Diamond, 1981: 9R). As discussed earlier, building these plants eventually cost about twice their original estimates, with the builders absorbing heavy losses.

LILCO expected that Shoreham's construction, licensing, and operation would proceed relatively smoothly, and that it would be producing power commercially in about four and one half years. Early, smaller nuclear plants had frequently been completed behind schedule and over budget, but not so much so as to dampen the utilities' enthusiasm for nuclear power. According to Arthur D. Little, The Atomic Energy Commission's licensing procedures were "relatively well-defined." LILCO's executives believed at the time that previous regulatory changes had resulted in only about six to twelve-month delays in earlier plants. Also, in the mid-1960s the states were the main environmental regulators, and LILCO had experience with New York's agencies. LILCO at least had studied nuclear power for several years prior to the Shoreham decision, and so felt prepared to undertake the project (Arthur D. Little, 1984: 21-22).

Indeed, contemporary reactor manufacturers, industry, and government suggested confidently that any remaining technical or economic problems could be solved without too much difficulty. As discussed earlier, even the critical reviews of nuclear power's costs, such as those of Philip Sporn, did not disagree too much with the optimistic views, and disagreed not by pointing out the flaws of nuclear technology, but by citing the advances of its fossil fuel competitors.

Basic Design of the Plant

The Shoreham plant uses nuclear fuel to boil water to generate steam which turns an electricity-generating turbine; thus, it is called a boiling water reactor. Radioactive fuel pellets are bundled in the reactor's core so that a controlled process of nuclear fission occurs. The pellets are placed in fuel rods which have walls (cladding) made of zircaloy. The rods are assembled in bundles and configured into the core of the reactor.

In a boiling water reactor circulating water removes the heat of fission. When the reactor's control rods are removed, the fuel reaches temperatures as high as 2,000 degrees Fahrenheit. The fuel cladding is designed to retain radioactive fission products and keep them from contaminating the circulating water. This cooling water is kept at a temperature and pressure such that it boils to form steam in the top of the reactor. The steam then travels through large pipes to the turbine, turning its blades to generate electricity. The spent steam continues to a condenser, where it rushes over thousands of small pipes containing cool water from Long Island Sound. Heat is transferred from the steam to the Sound water. Reactor water is then piped back to the pressure vessel and the cycle is repeated.

Control rods are inserted into the core of the reactor when it is necessary to stop the heat generating process, for refueling, or as a preventive measure. The control rods absorb the neutrons which initiate the radioactive process. Even after the control rods are in place, cooling water must be circulated past the core to remove residual heat. For a short time after the reactor shuts down, the spent fuel produces about six to seven percent of the heat that the plant would generate in full operation. The Shoreham reactor's temperature should drop to 125 degrees Fahrenheit twenty hours after the control rods are inserted.

Surrounding the reactor are steel-reinforced concrete walls and a steel liner backed up with additional reinforced concrete. The containment is a major part of the reactor's safety system, and has two parts. The top half, which contains the reactor, is called the drywell. The bottom half, called the wetwell, is a chamber partly filled with water. The drywell and the wetwell are separated by a concrete barrier, but vertical pipes, called downcomers, go through he barrier. If one of the reactor's steam lines were to break, the drywell would fill with high-pressure steam, which would take the path of least resistance down the downcomers and into the water pool. This "pressure suppression" pool would ease the pressure by turning the steam to water.

Several additional systems are provided for emergency core cooling in the event of a reactor accident. Also, the plant has emergency power generators so that pumps for cooling water and other safety systems can be operated in case of a normal or accidental power shutdown (New York State Fact-Finding Panel, 1983: 3[a]20-30).

Key Early Decisions

When the Shoreham project was started LILCO had to select a manufacturer and specific model of the reactor, and design an organization to build the plant. In later years, as the project's delays and cost overruns grew, LILCO's decisions on both counts were faulted. In retrospect, its early decisions were made partly for good (or at least adequate) reasons, and partly because of comfortable familiarity with certain contractors, sunk costs, and factors that--under norms of rational decisionmaking--should not "count."

LILCO's Selection of a Reactor and Containment System

In 1965 four companies had experience manufacturing nuclear reactors of 150 MW or more in the United States--General Electric, Westinghouse, Babcock and Wilcox, and Combustion Engineering. Of the four, General Electric (GE) and Westinghouse had the most experience. LILCO obtained bids for the Shoreham plant from GE and Westinghouse, suggesting that Babcock and Wilcox and Combustion Engineering were too new to the field. The two bidders estimated costs for the nuclear supply steam system, a steam turbine generator, and the initial fuel core.

LILCO retained Stone and Webster and S.M. Stoller Associates to evaluate the bids, and they concluded that the General Electric reactor system would cost $14.8 million compared to Westinghouse's $16.5 million. LILCO selected General Electric as the manufacturer of the nuclear generator and containment system.

The containment system is the barrier and equipment designed to prevent an accidental release of radioactivity to the environment. At the time General Electric's containment system design was called the Mark I system, and the initial GE bid for the Shoreham contract was made with the Mark I design. The Mark I had a large volume of air around the reactor, a water pool underneath the reactor, and pipes connecting the two. As discussed above, in the event of an accident, both air and water could relieve steam pressure. The Mark I, however, required large steel walls and equipment to absorb the pressure shock of an accident.

In 1968 General Electric, along with Stone and Webster, was developing a new type of containment system called the Mark II. The new system had larger walls and air volumes to absorb pressure than the Mark I. Thus, the walls of the containment system could be concrete rather than steel, and related equipment did not have to be as strong as in the Mark I. The Atomic Energy Commission reviewed the Mark II system in 1968 and commented favorably on it. LILCO had cooperated with GE in developing the Mark II, and suggested that the system provided greater protection from the dangers of excess steam pressure. Also, the Mark II would allow more flexible scheduling of construction, reduce the need for skilled crafts workers, and generally reduce costs. General Electric, Stone and Webster, and LILCO announced the Mark II system at the American Power Conference in 1968, and the design was selected as the containment system for Shoreham. Subsequently the Mark II system was adopted for use at the LaSalle 1 and 2 nuclear plants, Susquehanna 1 and 2, Limerick 1 and 2, Zimmer, WPPSS 2, and two plants in Mexico. In 1984 a report by Arthur D. Little, Inc.--commissioned by LILCO--commented that "In our opinion, LILCO's choice of Mark II containment was a reasonable, and based on evidence then available, low-risk decision to obtain advantages of evolving technology" (Arthur D. Little, 1984: 34).

In 1975 General Electric found serious design problems during testing of the Mark III system, problems which applied as well to the Mark II (Komanoff, 1981: 53). The pressures within the system were greater than originally estimated, and utilities had to spend tens of millions of dollars reanalyzing the forces on their plants' walls and equipment, and strengthening or replacing stressed walls and equipment. The impact on Shoreham was particularly serious because the water pool was much smaller than at other sites, and a smaller pool meant greater pressure. As will be discussed below, in 1968 LILCO had decided to increase the generating capacity of the planned facility from 540 to 820 megawatts, but did not enlarge the size of the reactor building itself. Thus, Shoreham's pressure suppression pool was relatively small given its generating capacity. For example, Shoreham's pressure suppression pool was built to hold 81,350 gallons of water, while the Mark II pool at the comparable Zimmer plant held 102,000 gallons. In an accident the pressure on Shoreham's walls and equipment would range from 27 to 97 percent greater than other Mark II's, all else being equal (Diamond, 1981: 10R).

The Decision to Use Stone and Webster as Construction Manager and Architect/Engineer

Utilities building nuclear plants had to decide whether to manage the construction themselves, or, if they lacked the necessary experience, to contract the management of construction out to another organization. Nuclear and fossil fuel plants differed enough for most utilities to contract the management of nuclear construction out to other firms.

LILCO had managed the construction of its fossil-fuel fired power plants, but in the 1960s lacked the experience and managers to step confidently into a nuclear project unassisted. Thus, it opted to hire a construction manager for the Shoreham plant. The Report by Arthur D. Little noted that "In the late 60s, LILCO was among relatively few utilities that constructed (built with their own work force) or managed construction (work done by several specialized contractors) of fossil plants. At the end of 1966, only three or four utilities with similar capabilities had decided to assume this role with the nuclear plants they were ordering. And these were utilities much larger than LILCO--Duke Power Company, TVA, and Pacific Gas and Electric. Moreover, at least one of these employed a professional firm as part of their team" (1984: 25).

In 1967 LILCO evaluated proposals to serve as construction manager from five firms--Bechtel Corporation, Burns and Roe, Ebasco, Stone and Webster Engineering Corporation, and United Engineers and

Construction. These five firms had designed or were designing ninety percent of the nuclear capacity that had been ordered in the United States (Arthur D. Little, 1984: 26). A committee composed of LILCO's Engineering, Production, Construction, and Purchasing departments recommended unanimously that Stone and Webster be retained as architect-engineer (A/E) and construction manager (CM). Stone and Webster's bid was about twelve percent higher than the lowest bidder. However, according to Arthur D. Little, the committee recommended that LILCO select Stone and Webster for several reasons.

LILCO had retained Stone and Webster to evaluate both the siting for the Shoreham plant and the GE and Westinghouse nuclear containment system bids, and was satisfied with its work. Stone and Webster also was familiar with the foundation problems likely to be encountered at Shoreham because it had construction experience at nearby Brookhaven National Laboratories.

Stone and Webster's professional staff also was relatively stable. Its key engineers averaged eighteen years of experience, with an average of eleven years of experience with Stone and Webster. It had sufficient numbers of personnel, and its Boston headquarters were less than 200 miles from Shoreham. Stone and Webster had extensive experience in nuclear construction management, from the first plant at Shippingport to current projects in New York State. It ranked third in the industry in experience in designing nuclear capacity, and LILCO said that it had been "successful in designing and constructing nuclear plants such as Yankee Rowe and Connecticut Yankee on schedule at or below budget estimates" (Arthur D. Little, 1984: 27). Stone and Webster had a very strong reputation in the 1960s. A former commissioner of the Nuclear Regulatory Commission commented that "in those days Stone and Webster had the reputation of being the outstanding nuclear architect-engineer" (Fitzgerald and Zorpette, 1987: 26).

The contract required Stone and Webster to furnish services consistent with the "best engineering and architectural practices." Stone and Webster was to design the basic characteristics of the plant. It would develop and implement Critical Path Method (CPM) and Planning, Evaluation, and Review Technique (PERT) scheduling for the stages of construction, and also budgeting and quality controls. Stone and Webster would prepare reports and testimony to the Atomic Energy Commission and other bodies. It would prepare specifications for bids for subcontractors, evaluate the bids received, and review and approve construction plans.

LILCO reserved the right to review and comment on the plant's designs before they were finalized, and to actually select the successful bidders for specific contracts. The company also retained purchasing and procurement functions in-house rather than delegate them to Stone and Webster.

LILCO maintained that, while it did not have experience managing the construction of a nuclear power plant, it did have extensive experience in purchasing and procurement activities and so could carry out that part of the project (Arthur D. Little, 1984: 29-32).

Some observers, particularly the Staff of the New York State Department of Public Service (DPS), later criticized the process by which LILCO selected Stone and Webster both for its early planning work on Shoreham and for the architect-engineer and construction manager contracts. LILCO did not seek competitive bids when selecting Stone and Webster for the earlier evaluation of the Shoreham site or the development of specifications for the nuclear steam supply system. LILCO also did not obtain its Board of Directors approval for these early contracts, violating the Board's policy on contracts (New York State Department of Public Service Staff, 1984: 76 [subsequently cited as Department of Public Service Staff]).

LILCO argued that there was a great demand for nuclear construction experience in the mid-1960s--recall that this was the time of the "great bandwagon market"--and so it was not possible to request fully competitive bidding for the Shoreham contract. However, its critics maintained that LILCO had followed the development of nuclear power for several years, and should have planned to obtain these services well in advance of needing them. The DPS Staff also contended that LILCO did not evaluate how well it would be able to work with Stone and Webster in the construction of the Shoreham plant:

> There was absolutely no consideration of the 'fit' between LILCO and other prospective A/E's and CM's [architect/engineers and construction managers]. It was important to identify the A/E and CM that would best fit the utility's organizational needs and the utility's philosophy of managing the plant's engineering, design, and construction. This would be true for any plant owner, but in LILCO's case the problem of ensuring compatibility was especially important because of LILCO's expressed intention to become involved in the construction of the plant. A reasonable manager would have evaluated his utility's capabilities, its needs and management philosophy specifically with a view toward selecting the A/E and CM whose resources would best complement the utility's. LILCO, having failed to discharge that threshold responsibility, could not know whether S&W or another of the major A/E firms would be LILCO's optimal partner or whether the A/E and CM should be separate firms (Department of Public Service Staff, 1984: 77).

These were potential managerial troubles. However, LILCO's critics also suggested that the ties between the two firms were tainted by conflicts of interest, for LILCO already had business and personal ties to Stone and Webster. Whitney Stone, Stone and Webster's co-owner, was a director of Chase Manhattan Bank, LILCO's largest stockholder. Chase Manhattan had extended a $5 million line of credit to LILCO. Stone and Webster owned eight percent of LILCO's major natural gas supplier. Eben Pyne, a member of LILCO's Board of Directors, also sat on the Board of Directors of W.R. Grace and Co., while the Grace chairman, J. Peter Grace, was a director of Stone and Webster (Diamond, 1981: 17R).

Similarly, the Executive Vice-President for LILCO who recommended that Stone and Webster receive the environmental site survey and A/E and CM contracts was the father-in-law of the Stone and Webster employee responsible for selling the services. Although the Department of Public Service staff suggested that this relationship might have biased the choice in favor of Stone and Webster, it lacked hard evidence to that effect (p. 78). While the utility maintained that the ties were "absolutely... not a factor" in the selection of Stone and Webster (Diamond, 1981: 17R), LILCO's critics said that the ties could have biased the selection for the architect-engineer and construction management contracts, and, at best, reflected poorly on the company's judgment.

However, of the many criticisms of LILCO's and Stone and Webster's performance at Shoreham, the arguments regarding conflicts of interest were made only in passing. The arguments that were made repeatedly, and most convincingly, were those regarding managerial problems.

Over time Stone and Webster's plants were recognized as being distinctively expensive. Stone and Webster built four of the five plants that, in 1986, stood as the most expensive in the nation, and only one of these was actually operating (Table 9). The plants were expensive partly because they were built in high wage areas, and partly because of other inherent project-related factors. Even allowing for these differences, however, Stone and Webster's plants were costly. Utility analyst Charles Komanoff described this as the "Stone and Webster effect," saying that "I can't explain it except to say that Stone and Webster in the 1980's has had a predilection to build expensive plants" (*Boston Globe*, July 8, 1986: 35).

Although remarkable cost overruns and delays eventually distinguished the Shoreham project, there is little in the record of the Public Service Commission's investigation of Shoreham or other documents to lead one to conclude that LILCO was being unreasonable in selecting Stone and Webster as a construction manager. Retrospectively we know that the Shoreham project damaged both LILCO and Stone and Webster. However, when LILCO selected Stone and Webster as a construction manager the decision seemed to be responsible.

Table 9. Comparative Costs of Selected Nuclear Plants (In 1984 Dollars)

Plant	Cost Per Kilowatt
Shoreham*	$4020
Nine Mile Point 2*	3948
Beaver Valley 2*	3819
Clinton	3693
River Bend*	3663
Hope Creek	3577
Seabrook 1	3153
Millstone 3*	2706
Summer	1630
Catawba	1235
Average Cost of 24 plants**	$2200

*Stone and Webster plants

**24 non-Stone and Webster plants completed in 1983 and beyond

Source: *Boston Globe*, "The Stone and Webster Effect," July 8, 1986.

The Decision to Increase Shoreham's Generating Capacity

The plans for Shoreham in 1966, based on the projected demand for electricity, called for a 540 MW facility. In the Summer of 1968 Long Island's peak electricity use increased 19 percent, twice LILCO's forecast of 9.5 percent. LILCO then reassessed its need for generating capacity, and a review of the Shoreham project was part of the reassessment.

Some company officials already were looking closely at the project's status independently of the unexpected rise in power demand. In 1968 Arthur Sugden, LILCO's Vice-President for Engineering, recommended to LILCO's Executive Vice President Edward Duffy that the company defer the construction of Shoreham. Sugden pointed out that General Electric had withdrawn from offering turnkey projects in the Fall of 1966, and thus that LILCO would have to absorb any cost overruns. This was disquieting because in its first two years the project's expected costs already had increased from $70 million to $170 million. Sugden argued that LILCO ought to evaluate whether such cost increases would continue and, if so, that LILCO might consider dropping its nuclear plant in favor of adding more fossil fuel capacity.

Further cost increases were in fact likely, according to Sugden's memorandum. The Atomic Energy Commission had established new regulatory controls on radioactive gases and construction requirements relating to seismic activity. Since 1966 the time required to obtain a construction permit from the AEC had increased, and LILCO had not yet obtained such a permit; thus, it probably faced a long proceeding to obtain the AEC's authorization to build Shoreham. Furthermore, there was a chance that the AEC would require nuclear plants to retrofit safety requirements, while there was little chance that fossil fuel plants would have to retrofit additional equipment. Technical problems, labor unions' demands for additional wages for work on nuclear sites, and environmental groups were also affecting nuclear plant construction. Because all these factors could increase Shoreham's costs, Sugden argued that LILCO ought to defer the project pending a reassessment of its needs for additional generating capacity, and also until it comparatively analyzed the current relative costs of nuclear versus fossil fuel-fired plants (Sugden report to Duffy, LILCO, October 1, 1968).

Up to this time Stone and Webster periodically had been analyzing the costs and benefits of Shoreham. On October 30, 1968, LILCO's Chairman of the Board John Touhy told the Board of Directors that Stone and Webster's most recent analysis concluded that a 540 MW nuclear plant--such as Shoreham--now appeared to be more expensive than an equivalent fossil fuel unit. The analysis reviewed cost and schedule overruns for nuclear plants at other sites, including the Oyster Creek and Nine Mile 1 projects, recent increases in construction labor wages, and decreases in productivity in nuclear construction. The analysis also reviewed the unexpectedly high peak electricity demand in the Summer of 1968, and suggested that LILCO would probably have to add additional generating capacity before 1973, the scheduled completion date for Shoreham. The Board tentatively decided to postpone the Shoreham Project, and authorized a 380 MW fossil unit to be built at Northport for operation in May 1972. LILCO then determined that the logical start-up date for its next large generation unit (fossil or nuclear) would be May 1975. Approximately 850 MW of generating capacity would be needed at that time (Department of Public Service Staff, 1984: 35).

An alternative to postponing Shoreham and increasing fossil fuel capacity was to increase the size of the Shoreham plant itself. That possibility was discussed at the October 30th Board meeting. Chairman of the Board Touhy indicated that no final decision on increasing the size of the Shoreham reactor had been made, although preliminary site work on the plant would continue. However, in a letter dated November 6, 1968 (one week after the Board meeting), a Stone and Webster Vice-President told LILCO President Duffy that Stone and Webster had stopped work on the 540 MW unit, and thanked Duffy for his "authorization for the design and management of construction of an 820 MW unit for operation in 1975" (letter from W.F. Allen to E.C. Duffy, November 6, 1968).

Thus, LILCO's management apparently authorized Stone and Webster to build an 820 rather than 540 MW unit five months before the Board of Directors actually approved the upsizing in March 1969 (Department of Public Service Staff, 1984: 59). While LILCO later argued that its management only approved preliminary work necessary for planning cost estimates, the DPS Staff responded that "The S&W letter makes it clear that Mr. Duffy gave blanket approval to proceed with engineering and construction management for the larger unit" (p. 59).

In February 1969--four months after Stone and Webster apparently received authorization to proceed with the 820 MW unit--LILCO issued a report projecting nuclear versus oil-fired plant costs for 1975. LILCO said that an 820 MW nuclear plant would have an advantage over an oil fired plant if the cost of residual fuel oil exceeded 26.3 cents/million BTU's (1969 dollars). The prevailing cost of fuel oil was 27 cents/million BTU, and the cost was expected to rise more rapidly than the cost of nuclear fuel. The Report labeled the nuclear option as superior because of these relative fuel costs.

Also, LILCO maintained, the fossil fuel unit option had a possible regulatory disadvantage. A new fossil fuel plant probably would be sited at LILCO's Northport facility. Under U.S. Environmental Protection Agency regulations, building a new fossil fuel unit at Northport might cause the entire facility there, including existing units, to come under new environmental regulations, increasing fuel oil costs still further. (This factor was not included in the cost calculations).

Preliminary work and applications assumed Shoreham to be a 540 MW unit. Stone and Webster told LILCO that this preliminary work could be applied to the 820 MW unit, and that the AEC had agreed to let LILCO amend its 540 MW license application and maintain its "place in line" for construction permit hearings. Suspending work on Shoreham and picking up work on a nuclear plant at a later time would cost LILCO its priority for construction permit hearings (Arthur D. Little, 1984: 36). The application for the construction permit would have to be re-written and refiled for the larger unit, however.

On March 26, 1969, the LILCO Board of Directors approved the proposal to increase Shoreham's capacity to 820 MW and to proceed with the project. A critical element of the decision was a belief that the project could be upgraded from a 540 to an 820 MW plant without major alterations in site preparation and other work. The plans for reactor building size were not altered; thus, an 820 MW plant was eventually fitted into a space intended for a 540 MW system. As discussed later, this decision substantially increased the difficulty of building Shoreham.

Summary

The Federal government strongly promoted the growth of atomic energy in the twenty years up to 1966. Utilities and other manufacturing firms eventually began to invest heavily in nuclear power plants, and, with the Atomic Energy Commission, regarded and publicized nuclear power as the most desirable future source of electricity. LILCO, like other utilities, decided to undertake nuclear construction programs in the great surge of new orders in the mid-1960s.

LILCO was a sophisticated, successful utility with a large amount of experience in fossil fuel plant construction and operation. Although many utilities had studied nuclear power plants, most had little experience actually building them. This was the case with LILCO. Thus, when it made the key early decisions regarding Shoreham--the decisions to install the Mark II reactor, to use Stone and Webster as a construction manager/architect engineer, and to upgrade the system from 540 to 820 MW--it was operating in unfamiliar territory.

Critics traced many of Shoreham's future problems to Stone and Webster, the Mark II, and the consequences of "jamming" an 820 MW system into a space designed for a smaller system. Indeed, LILCO acknowledged, retrospectively, many of these problems. LILCO and Stone and Webster did not thoroughly analyze the decision to upgrade the size of the plant; LILCO admitted later that it should have built the original 540 MW unit, adding an identical unit later to accomodate increased demand, rather than upgrading the plant (Diamond, 1981: 10R). LILCO complained of substandard performance by Stone and Webster. (Although it selected Stone and Webster for its proposed Jamesport project, preferring "the devil we knew" to hiring another company [Stone and Webster memorandum from E.P. Donegan to N.B. Cleveland, September 24, 1973].) However, with the exception of the upsizing issue, the decisions seemed responsible at the time they were made. Only when these decisions interacted with Shoreham's future political and organizational problems did they play a large part in the project's difficulties. The next Chapter discusses the early politics of Shoreham.

Appendix to Chapter 2
Chronology of Key Events

1965

April 13: LILCO announces plans to build the Shoreham nuclear plant, at a cost of $65-75 million.

1966

August 16: LILCO's Board of Directors approves the selection of Stone and Wester for the environmental site survey at Shoreham.

1967

January 1: General Electric and Westinghouse abandon turnkey nuclear projects.

February 21: LILCO's Board of Directors authorizes the contract with General Electric to supply the turbine generator, nuclear steam supply system, and initial fuel load for Shoreham.

May 23: Edward Duffy, LILCO's executive vice-president, recommends the selection of Stone and Webster as architect/engineer and construction manager for the Shoreham project.

July 26: LILCO's Board of Directors approves the construction of Shoreham Unit 1, a 540 MW nuclear unit for commercial operation in the Spring of 1973 at a cost of $124 million.

October 9: LILCO announces plans for a nuclear facility at Lloyd Harbor.

1968

May 15: LILCO submits the Preliminary Safety Analysis Report (PSAR) for a 540 MW reactor at Shoreham to the Atomic Safety Commission.

October 1: LILCO vice-president Arthur Sugden recommends that the Long Island Lighting Company defer the Shoreham project in light of emerging concerns about nuclear plant construction.

October 30: LILCO Board Chairman Charles Touhy advises Board of Directors that recent Stone and Webster analysis has concluded that a 540 MW nuclear plant would be more expensive than an equivalent fossil fuel unit. Board decides tentatively to postpone Shoreham project, and authorizes a 380 MW fossil fuel unit to be built at Northport.

November 6: Stone and Webster Vice-President advises LILCO President Duffy that Stone and Webster had stopped work on the 540 MW unit, and thanks him for "authorization for the design and management of construction of an 820 MW unit" at Shoreham.

November 21: LILCO notifies the Atomic Energy Commission that it plans to construct an 820 MW unit at Shoreham with a commercial operation date of May 1975. The Atomic Energy Commission agreed to retain the same docket for the amended PSAR for the 800 MW Shoreham unit.

1969

February: LILCO issues a report saying that an 820 MW nuclear plant currently would have an economic advantage over an oil-fired plant.

March 26: LILCO's Board of Directors authorizes the construction of an 820 MW unit at Shoreham at an estimated cost of $217 million, with a commercial operation date of May 1975.

April 21: LILCO files amended Preliminary Safety Analysis Report to reflect 820 MW Shoreham unit.

June: Arthur Wofford hired and named LILCO project manager for Shoreham.

July 1: Site clearing and grading begins at the Shoreham site.

September: LILCO increases estimated cost for Shoreham to $261 million, with projected commercial operation in May 1975.

Chapter 3

THE EARLY POLITICS OF SHOREHAM

The first local reaction to LILCO's proposal to build Shoreham was favorable. But a movement against nuclear power on Long Island shortly emerged, first forming outside of the Shoreham area. Partly by design and partly through chance, the anti-Shoreham movement delayed the project at critical early points, with lasting consequences.

The Contrasting Reactions to Shoreham and Lloyd Harbor

In January, 1966 LILCO applied to the Town of Brookhaven for a zoning permit change which would allow it to build Shoreham, pending approval from Federal and State agencies. The local reaction was favorable. On January 21, 1966, the president of the Wading River Civic Association wrote to the LILCO Board Chairman that "...we realize that your company, as a public utility, is very sensitive to public reaction to your plants...having [an atomic plant] as a neighbour has produced a most favorable reaction in our community" (Diamond, 1981: 12R). At a Brookhaven Town Board hearing on the zoning permit in June virtually all speakers supported the plant. The Vice-President of the Shoreham Civic Association said his group favored an atomic plant but opposed a coal plant. The Riverhead Supervisor called the plant "a perfect type of industry for the area." The Suffolk County Executive supported the plant, and the Brookhaven Town Supervisor said that he was "more than pleased" with LILCO's project. The *Port Jefferson Record* said that the plant would have "enormous" benefits, including millions of dollars in school taxes. Later, in 1977, it was estimated that adding Shoreham to the tax base would leave $200,000 in assessed property value behind every pupil in the Shoreham-Wading River School District, making it the richest school district in New York State. Scientists at the Brookhaven National Laboratory, which had done much developmental work on nuclear power, also supported the plant. The Town Board voted unanimously to give Shoreham a zoning permit.

On October 9, 1967 LILCO announced that it planned to build another nuclear power plant at Lloyd Harbor, New York. Lloyd Harbor was a residential upper class community near the Nassau-Suffolk County line. Mayor George Barclay wrote a letter to the Village's 750 taxpayers stating that the $150 million plant would provide "massive" tax relief. LILCO said that the plant "would be designed to blend in with the land...with nothing showing above the treeline. Any supplies would be brought in by way of Long Island Sound, and nonradioactive waste would be dumped far out into the water" (*New York Times*, October 10, 1967: 31).

The proposed Lloyd Harbor plant was not as well received as the Shoreham proposal. As noted above, Lloyd Harbor was an upper-class community. Ronald Rasch, a retired banker who was chairman of the group Action for the Conservation and Preservation of the North Shore, said that the group was opposed to the general principle of development in the area because it contained "some of Long Island's most valuable remaining wildlife, marshes, and lakes." The proposal prompted the founding of the Lloyd Harbor Study Group, which eventually emerged as the most effective early opponent of not only the Lloyd Harbor plant but also of Shoreham. According to Ann Carl, one of the Group's founders, Lloyd Harbor

> [P]rided itself on its attention to conservation: an ordinance protected wetlands from destruction; pesticides were restricted; ponds, shoreline, fish and wildlife were protected. Living in the community were the president of the Long Island chapter of the Nature Conservancy, the chairman of the Town Conservation Advisory Council, and officials of the Huntington Audubon Society and several other strong local and national conservation action groups. Most of the residents were technical people: engineers, physicians, businessmen in technical fields. The Long Island Lighting Company chose the Town of Huntington community for its second nuclear plant proposal. It was a community which already knew something of the growing worldwide controversy over nuclear plants. It was not ready to believe unquestioningly LILCO's assurance of benefits from this neighbor in a residential community (Carl, 1972: 32).

In November 1969, the Lloyd Harbor Study Group released a report charging that the proposed Lloyd Harbor plant would be "unsafe, unreliable, and unclean" (*New York Times*, November 9, 1969: 27). LILCO was surprised by the resistance. A LILCO attorney, Edward Barrett, later commented that "Maybe we were naive. But we believed nuclear was the best alternative; we thought it would be a nice, clean neighbor," and that, short of scrapping the plant, "we were willing to accomodate." Former LILCO President Uhl said that "in those days, siting a nuclear power plant wasn't perceived as being a huge problem...It was a nice clean plant...a beautiful thing for any community" (Diamond, 1981: 12R). Facing organized opposition, LILCO abandoned the Lloyd Harbor project. But the Lloyd Harbor Study Group then turned its attention to Shoreham.

The Purpose of AEC/NRC Hearings

By 1970 LILCO had finished its application to the Atomic Energy Commission for Shoreham's construction permit. The hearings on the application began in September 1970. Previously, AEC construction permit hearings had lasted no longer than two weeks. LILCO hoped, and expected, that the hearings would not extend much beyond that time, with a construction permit issued shortly thereafter. The construction permit hearings actually finished about two and one half years later, and the construction permit was issued in April 1973. Shoreham's political, regulatory, and managerial problems grew rapidly in this period.

The national push for nuclear power in the 1950s and 1960s extended the requirements for public participation in decisions on nuclear power. The 1954 Atomic Energy Act required extensive licensing and regulatory procedures for the construction and operation of nuclear power plants. The 1957 amendments added a requirement that the Atomic Energy Commission hold hearings when a utility applied both for a construction permit as well as for an operating license. In 1962 the requirement that a hearing take place at the operating license stage was dropped to expedite reactor licensing; the AEC maintained that its procedures still allowed for adequate public scrutiny of operating license decisions. How effective these procedures were, or how faithfully followed by the Atomic Energy Commission, were contested points. In the 1960s some actors argued that the requirements were neither effective nor followed. Others, favoring nuclear power, said that the spreading requirements were preventing the industry's development.

Up to the early 1960s, groups opposed nuclear power only in intermittent and ad hoc ways. There were exceptions, such as opposition to a plant at Bodega Bay in California and in the Ravenswood section of Queens. In general, however, the opposition consisted of relatively noncontentious questioning of utilities or the AEC in town meeting settings (Del Sesto, 1979: 119-120).

By the middle of the 1960s the growth of nuclear power was more conspicuous--this was, after all, the period of the "Bandwagon Market," with fifty-one orders placed in 1966-67--and thus more likely to be opposed. In 1967 the AEC estimated that 23 to 30 percent of electricity in the United States would be generated by nuclear power by 1980, and 50 percent by the year 2000 (U.S. Atomic Energy Commission, 1968: 6). The President's Office of Science and Technology estimated in 1968 that almost 500 new electrical generating stations would be needed by 1990 to keep up with demands for electricity (United States Office of Science and Technology, 1968: Chapter 1).

Also, the typical *size* of the reactors was increasing. The WASH-740 study in 1957 had forecast low probability but potentially dire consequences from nuclear accidents. Since then larger reactors, with potentially larger hazards, had either been built or had been ordered. Critics argued that the AEC, the Congressional Joint Committee on Atomic Energy (JCAE), the utilities, and the Department of Defense were fostering nuclear power largely outside political controls. For example, in 1972 Harold Green wrote that

> The political structure of atomic energy is in reality a closed loop. Starting with the premise that atomic technology should be developed, introduced, and used, the JCAE and the AEC have usually cooperated symbiotically to resolve policy issues behind the scenes before legislative proposals can reach public view. By the time these proposals emerge, it can be said that they are "noncontroversial," and they usually breeze automatically through Congress. Benefits of nuclear technology have usually been presumed, while risks have been ignored or dismissed in boldly optimistic terms (1972: 76-77).

By the end of the 1960s plans to open nuclear power plants around populated areas were far more likely to be controversial and contested.

The Licensing Process

In 1970, in the first stage of the licensing process, the applicant utility would meet with State and Federal officials. The applicant had to file a Preliminary Safety Analysis Report (PSAR) which analysed the proposed site and reactor; operations and special characteristics of the proposed plant; possible unusual events (like accidents) and procedures for dealing with them; geographic data; and procedures for handling wastes. The AEC's Office of Regulatory Activities evaluated the PSAR, consulting with other agencies if it wished. The applicant would be asked to respond to any criticisms of the PSAR or the plant itself. After providing additional information the applicant and the AEC would informally negotiate changes in the plant or the PSAR until disagreements had been eased as much as possible.

The AEC would then prepare a "Safety Evaluation Report." The Advisory Committee on Reactor Safeguards (ACRS) would review the Safety Evaluation Report before it was released to the public. The ACRS had fifteen members, representing various scientific disciplines. It had been established in 1953 as a subunit of the Atomic Energy Commission, but the 1957 amendments to the Atomic Energy Act made the ACRS an independent body. The ACRS was responsible for advising the AEC on construction permits for nuclear plants, analyzing the relative degree of safety of the proposed plant, and suggesting where its safety could be reasonably improved.

A major change in the licensing process emerged in September 1971, when applicants were required to file full environmental reports when applying for construction or operating license permits. Prior to that time the Atomic Energy Commission maintained that it had to consider radiation hazards *only* when reviewing construction and operating license applications. In July 1971, the Federal Court of Appeals for the District of Columbia, deciding against the AEC in a case involving the Calvert Cliffs nuclear plant under construction near Maryland's Chesapeake Bay, called this position a "mockery" of the National Environmental Policy Act of 1970 (*Calvert Cliffs Coordinating Committee v. United States Atomic Energy Commission*, 449 F. 2d 1109 [D.C. Cir., 1971], p. 1117). In September, the AEC, which did not appeal the decision, ordered full environmental reviews of ninety-six reactors for which it had granted construction permits or operating licenses after passage of the Environmental Policy Act on January 1, 1970. Current construction permit or operating license hearings also now would have to hold full environmental reviews, bringing into play a new array of arguments and actors. Shoreham was one such plant.

The environmental review had to evaluate the technical, economic, and environmental costs and benefits of the facility, and alternatives to the construction and operation of the plant. The AEC staff would then circulate a "Draft Detailed Statement of Environmental Considerations" for comment by other governmental agencies. After receiving comments the AEC staff would issue a "Final Detailed Statement of Environmental Considerations" to be sent to the Council on Environmental Quality. The public hearing on the construction permit would then begin.

Until 1962 the hearing examiner was generally a lawyer, leading to criticisms that the chief presiding office was not competent to evaluate the technical issues on the record. The Joint Committee on Atomic Energy recommended a change in the process, and the 1962 amendments to the Atomic Energy Act created the Atomic Safety and Licensing Boards (ASLB). The ASLBs had three members, "one of whom shall be qualified in the conduct of administrative proceedings and two of whom shall have such technical or other qualifications as the Commission deems appropriate to the issues to be decided." The board members included biologists, lawyers, nuclear scientists, and others from both inside and outside the AEC.

All participants in the hearings could give oral or written testimony, and cross-examine other participants. A number of issues, *including the AEC's regulations themselves*, were not subject to discussion. Critics contended that restricting the hearing to "narrow technical issues" prevented intervenors from raising many of the fundamental health-safety issues which motivated their involvement in the hearing.

After the hearing the ASLB issued a decision based on the record. Generally, this involved the issuing of a provisional construction permit called a Limited Work Authorization. This provisional permit was the basis of the final construction permit, although parties could file objections with the ASLB, as well as with the AEC and the Federal courts. After the Limited Work Authorization was issued utilities generally began site preparation for the plant, or construction at their own risk.

Once the final construction permit was issued full-scale construction began. Before the completed plant could operate there had to be a review for an operating license. This review included further technical review, description and analysis of the plant and its safety systems called a Final Safety Analysis Report (FSAR), and a final Environmental Statement. While a public hearing was not mandatory at this stage, if a party requested a hearing the AEC had to conduct one.

Del Sesto and others argued that public hearings generally had little impact on the decision whether to build the plant, on its features, or on its operations. The hearings occurred only after extensive technical reviews, consultations, and so forth among technical "insiders" such as the AEC, the utility, and other organizations. By the time the hearing occurred these parties would have resolved most issues in a way satisfactory to them.

> To be sure, considerable effort was expended well before the public hearing to demonstrate that the proposed facility was "in fact" sound and would not endanger the health and safety of the public. Thus, once the AEC's regulatory staff and the ACRS approved an application, it seemed that the eventual issuance of a construction permit remained a formality...the hearing seemed designed more to introduce and reaffirm decisions that had already been made rather than to allow the public a real voice in these decisions, and, hence, the appearance and not the reality of public participation was what was actually demonstrated (1979: 133-134).

The Shoreham construction permit hearings were the first to be held after passage of the National Environmental Policy Act (NEPA), which required agencies to fully consider environmental considerations, and after the Calvert Cliffs decision, which forced the Atomic Energy Commission to fully implement NEPA. The hearings are considered a central moment in the beginning of the anti-nuclear movement. Richard Pollack, a former head of Critical Mass, commented that "Even though a lot of people woke up after Three Mile Island, we regard Shoreham as the real beginning" (Diamond, 1981: 12R).

Shoreham's Hearings

The Shoreham construction permit hearings opened on September 21, 1970. At the opening session LILCO President Edward Duffy said that summertime power demand on Long Island had increased over 100 percent in ten years, and that between 1970 and 1975 the demand was projected to almost double once again from 2,156 MW to 4,000 MW. "Without Shoreham," he maintained, "there will be no reserves. That prospect is intolerable and against all sense and experience." A group of scientists from Brookhaven National Laboratories, having formed an organization called "Suffolk Scientists for Cleaner Power and Safer Environment," also spoke on behalf of the plant.

In contrast, Shoreham's opponents declared that they would make the hearings a comprehensive examination of the state of nuclear power. Irving Like, the chief attorney for the Lloyd Harbor Study group, said that "Our purpose is not so much to block a particular plant, but to encourage the nation's scientific community to come in and testify and make this a landmark case" (Diamond, 1981: 12R). In his opening statement Like said that the Lloyd Harbor study group would present several dozen witnesses, "a distinguished group from all parts of the nation including one Nobel Laureate." He criticized the organization of the AEC, saying that its role as promoter of nuclear power made it an ineffective regulator of nuclear power; thus, he argued, the Atomic Safety Licensing Board conducting the hearing had "a duty to lean over backward in making sure this is a full environmental hearing." Jack Campbell, an attorney who headed the ASLB for the Shoreham hearing (and a former Governor of New Mexico), reminded Like that the AEC felt it had only jurisdiction over the environmental effects of radiation, and thus could not, or would not, consider other environmental effects. (As noted earlier, this position was rejected ten months later in the *Calvert Cliffs* decision).

Like outlined a list of 17 "minimum requirements" that ought to be conditions for a construction permit, including the total elimination of any radiation emissions, elimination of thermal pollution, and a containment structure strong enough to withstand the crash of a large airplane or attacks from another nation. In later sessions the Lloyd Harbor Study group focussed on the possibility of catastrophic damage accidents related to the anti-aircraft missile base near Shoreham or the regional growth of air traffic (*New York Times*, September 22, 1970: 88; November 1, 1970: 78).

It soon became apparent that the hearings would last longer than anyone anticipated. Their length became a major issue starting in January 1971, when Irving Like delivered a paper to the American Bar Association on turning a nuclear reactor hearing into "a multi-media confrontation." Like said that "Each day must become a dramatic and suspenseful event...phrase statements in a lively manner--which will read or sound well when recorded in the next day's press or radio account." The next month LILCO introduced Like's paper as evidence that the Lloyd Harbor Group was more interested in delaying the hearings to block the plant than in examining its merits or problems (Diamond, 1981: 13R).

By January the number of sessions scheduled had increased from three a month to six a week. In late January the presiding ASLB announced that the hearings would extend beyond April 3, 1971. That date was critical, because on April 3 New York State's standards governing water discharges would go into effect. Under the new standards LILCO had to submit studies showing that the plant would not discharge water exceeding the temperature of the surrounding water of Long Island Sound by one degree in summer and four degrees in winter. The State would then conduct environmental hearings on the plant, in addition to the Federal hearings. Going through this additional set of hearings would delay the project for a minimum of one year.

In March 1971, several scientists, including Nobel Prize winner James Watson; Dr. Alice Stewart, the director of social medicine at Oxford University; and Dr. Benjamin Sonnenblick, a radiation biologist at Rutgers University and member of the New Jersey Commission on Radiation Protection testified on behalf of the Lloyd Harbor study group. In general terms Watson rejected the safety of atomic power, noting that "I'm not trusting the estimates I've seen. Remarkably powerful and big companies have ruined themselves and everyone else by faulty technology," and that "the amount of research now being done on the connection between cancer and radiation is totally inconsistent with proposals for widespread introduction of nuclear plants into highly populated areas." The *New York Times* noted that Alice Stewart's testimony focussed on a study revealing "a general cancer hazard associated with obstetric radiology [showing] that risk depends on the exposure and that there is no totally safe minimum exposure" (March 15, 1971: 24).

In April, 1971, Dr. Arthur Tamplin of the AEC's Lawrence Livermore Radiation Laboratory testified that the AEC's standards for radiation emissions ought to be lowered to one tenth of their current levels, as "Present radiation exposure guidelines are simply based upon the hope that the genetic effects won't be intolerable." The *New York Times* offered the following description of the testimony of Dr. Ernest Sternglass, professor of radiation health physics at the University of Pittsburgh:

> Dr. Sternglass said that low-level radiation "cannot only lead to significant rises in diseases previously known to be produced...such as congenital effects and cancer, but it also appears to act indirectly so as to produce small decreases in maturity at birth that in turn can increase the chance of early death from various causes such as respiratory and infectious diseases."

Dr. Sternglass was extensively cross-examined by Suffolk Scientists for Cleaner Power and a Safe Environment, which is composed mostly of scientists from the Brookhaven National Laboratory.

Extrapolating from vital statistics of infant mortality and occurrences of congenital defects and leukemia in certain states over a period of years before, during and after nuclear weapons tests, Dr. Sternglass said "Drastic downward revisions of isotope discharges and concentrations in the environment by 100 to 1,000 times must be required, unless we wish to see major catastrophes as the result of daily discharges of radioactivity from peaceful nuclear detonations or the operation of nuclear reactors and nuclear fuel processing plants."

Dr. Ferdinand Shore, professor of physics at Queens College and a member of the Suffolk Scientists group, interrogated Dr. Sternglass at length on his method of research and analysis, strongly challenging his use of statistical deviations (April 18, 1971: 60).

Professor Sternglass testified again on the association of radiation and infant mortality on May 11; he was supported at the hearing by Dr. Morris De Groot of the Statistics Department at Carnegie Mellon and Dr. Harold Rosenthal, professor of physiological chemistry at Washington University of St. Louis, and was critically cross-examined by scientists from the Brookhaven National Laboratory. The headline for the story in the *New York Times* was "Hearing on L.I. is Told Nuclear Reactors Appear Linked to Infant Mortality" (May 12, 1971: 24).

Two events in 1971 caused the hearings to be extended further. The first event was a major review of emergency core cooling systems at nuclear plants. In early 1971 experiments at the Atomic Energy Commission's National Reactor Testing Station in Idaho indicated that in a "loss of coolant" accident the reactor fuel would become hotter than previously believed. Containment systems were designed so that back-up cooling water would flow into the system. However, in an experimental simulation, back-up cooling water bypassed the overheating reactor core, discharging out of the same pipe break as the primary cooling water. Dr. Morris Rosen, the head of the AEC's Division of Reactor Standards, wrote an internal memo in 1971 saying that the performance of the emergency core cooling system (ECCS) could not "be defined with sufficient assurance to provide a clear basis for licensing" (Marcus, 1986: 9; Ford, 1982: 108).

In May 1971 the AEC began reviewing emergency core cooling systems. The review did not affect operations at current plants, or construction at those plants with construction permits. However, construction permit hearings under way, including Shoreham's, were suspended until the review was completed. In August the AEC's regulatory staff issued an analysis saying that the proposed Shoreham emergency core cooling system was adequate. Hearings focussing on the Shoreham ECCS began on November 4, 1971, and were concluded in the late Spring of 1972.

The second event further delaying hearings was the District of Columbia Court of Appeals' Calvert Cliffs decision in July, 1971. As discussed earlier, that decision ordered the Atomic Energy Commission to do a full environmental review. The AEC, which did not appeal the ruling, required all hearings under way (and all subsequent hearings) to include a full environmental review of the plant. In June, 1972 the AEC's 370-page "Draft Environmental Statement" for Shoreham concluded that "After weighing the environmental, economic, technical, and other benefits against environmental costs and considering available alternatives, it is concluded that the action called for is the issuance of a construction permit for the Shoreham nuclear power station" (*New York Times*, June 19, 1972: 37). Late in June the New York State Department of Environmental Conservation, which had supervised nine months of State hearings on the impact of the plant on water quality, also concluded that the plant could be built and operated without violating water quality standards.

However, hearings on the AEC's environmental review still had to be conducted. These began on December 4, 1972, and concluded on January 19, 1973. On April 12, 1973, the AEC announced that it was granting LILCO a construction permit for Shoreham. The permit restricted permissable radiation release in gas or liquid form, required monitoring of marine organisms and how they were affected by the discharge of heated water from the plant, and required protection of nearby wetlands and beach areas during construction. LILCO, which had been preparing the site for several years, announced that it would begin construction "immediately" by pouring concrete for the foundation of the reactor building. The construction permit was issued five years after LILCO had applied for the permit, and two and one-half years after the construction permit hearings began in September 1970.

Two Views of The Hearings

There were two views of the hearings. The view of the Lloyd Harbor Study Group and other opponents of the Shoreham project was that the hearing record "is one of the most complete compendiums available on the pros and cons of nuclear power at the present state of the art" (Carl, 1972: 32). According to this view, the AEC inevitably would have granted the construction permit because of its bias in favor of nuclear power. However, the hearings shattered the image of nuclear power as a well-understood, safe technology.

Stuart Diamond noted that, if the plant's critics could be accused of pursuing a "fishing expedition" in its lengthy cross-examination of personnel from LILCO, General Electric, and the Atomic Energy Commission, at least "the opponents could claim some catches" (1981: 13R). Some of Lloyd Harbor's most important arguments regarding flaws in tests, and breakdowns in reactor equipment and safeguards, were confirmed either by cross-examination of witnesses or by contemporary events such as the simulated failure of an emergency core cooling system at the AEC's test facility. A Lloyd Harbor Study Group leader argued that a series of "distinguished scientists...questioned the reasonableness of the AEC's emission standards and, in some cases, the vested interests represented in the makeup of the standard-setting bodies, the former Federal Radiation Council and the National Council on Radiation Protection and Measurements" (Carl, 1972: 34). She argued that the hearings showed scientists as being divided over nuclear power's safety, and that, until the safety questions were settled, it was not reasonable to make lasting commitments to nuclear power plants.

Another view of the hearings was that they functioned primarily as a stage for the Lloyd Harbor Study Group and other opponents of the plant to oppose Shoreham; that the opposition went well beyond reasoned argument about the plant, quickly crossing into an attempt to delay and thereby kill the plant; and that historical circumstances combined with these delaying tactics to set the Shoreham project back severely. Vance Sailor, a central organizer of the Brookhaven National Laboratory scientists favoring the plant, wrote that

> Once the hearings were underway, they were prolonged by the actions of LHSG [Lloyd Harbor Study Group] and soon broke all endurance records for hearings of this type. Between September 21, 1970 and May 19, 1971, 53 sessions were held of which more than 40 were devoted almost exclusively to LHSG "performances"--cross-examination, presentation of witnesses and/or legal argument. The first LILCO witnesses were sworn in and seated on September 22, 1970. LHSG began cross-examination of these witnesses on the morning of September 23, 1970. These same witnesses were still being interrogated by LHSG on December 16, 1970 when the [Atomic Safety Licensing] Board chairman, after several weeks warning, ruled that the questions had long since become excessively repetitious and that LHSG had been allowed enough time to complete its cross-examination. Cursory examination of the transcript will show that the same subject matter was covered again and again, despite the fact that satisfactory answers to all intelligible questions had been obtained repeatedly. Many questions could not be answered simply because they were not intelligible.

Sailor similarly criticized the Lloyd Harbor Study Group's witnesses themselves generally as having no credentials qualifying them to speak seriously to the technical issues regarding nuclear power. Sailor and others also criticized other admittedly qualified LHSG witnesses for presenting solely "general" testimony, pointing out that the witnesses acknowledged under cross-examination that they had not examined the specific materials in LILCO's Shoreham proposal. For example, Sailor noted that Nobel Prize winner James Watson's testimony was "philosophical and general in nature and did not include any data to support his conclusions. Cross-examination revealed he had not studied the Preliminary Safety Analysis Report, and did not know the magnitude of natural radiation background at the site of his own laboratory in Cold Spring Harbor. Watson's testimony was a disappointment to us. Despite the fact that he holds a position of high prestige as director of a distinguished laboratory and has earned international acclaim for his research in 1951-53 on the geometrical structure of DNA, he presented no data to support his comments, and there was no evidence that he had devoted serious study to the problem before testifying" (1972: 29).

Prolonging the hearings, Sailor and others argued, delayed construction of the plant by at least two years. The hearings' length also meant that Shoreham was "caught" by the *Calvert Cliffs* decision and the need for an extra review of the emergency core cooling system, with the extra environmental and safety reviews approving the plant's design. Thus, he maintained, the Lloyd Harbor Study Group Intervention, and the prolonged nature of the hearings, served no purpose other than as an effort to delay and block the plant.

Whether one sees the record-breaking Shoreham hearings as an historically important examination of nuclear power, or as a move to block the entire project, the delay did contribute to the plant's later problems. By the time the construction permit was issued the project faced a number of delay-related problems that required efficient management and, as discussed in the next chapter, neither LILCO nor Stone and Webster coped with the project particularly efficiently.

Furthermore, because of the intense opposition to the construction permit, the NRC expected intense opposition to the Shoreham *operating license* once the plant was completed. Thus, throughout the project the agency examined LILCO's regulatory compliance closely, and this probably increased the difficulty of managing the project. A former NRC official commented that

> ...The use of Regulatory Guides in the Shoreham proceeding demonstrates the way the Shoreham OL [Operating License] review was conducted that was different from other OL reviews in the same time frame. Regulatory Guides are not requirements, as I have indicated, but deviations from them need to be justified extensively. How extensively depends on staff attitudes. In Shoreham's case it was thought to be desirable to build the strongest technical record that would show equivalent conformance to this guidance. Shoreham was always held to a higher standard and this frequently resulted in delays that did not impact other plants.

The reason for this situation was straightforward. The legal staff, based on the [construction permit] experience, wanted the strongest technical case possible for the Shoreham OL hearing. This naturally effected [sic] the depth of review to which the application was subject (cited in New York State Public Service Commission, 1985: 85-86).

Shoreham's construction did not proceed smoothly. After Shoreham received a construction permit in 1973, it took about twelve years for the plant to be completed. When low-power testing started in 1985, Shoreham's cost per kilowatt hour was estimated to be by far the highest in the nation. The next Chapter discusses the project's management and regulatory problems during construction.

Appendix to Chapter 3
Chronology of Key Events

1965

April 13: LILCO announces plans to build the Shoreham nuclear plant, at a cost of $65-75 million.

1966

June: Brookhaven Town Board votes unanimously to issue zoning change to allow LILCO to build Shoreham.

1967

October 9: LILCO announces plans for a nuclear facility at Lloyd Harbor.

1968

May 15: LILCO submits the Preliminary Safety Analysis Report (PSAR) for a 540 MW reactor at Shoreham to the Atomic Energy Commission.

1969

April 21: LILCO files amended Preliminary Safety Analysis Report to reflect 820 MW Shoreham unit.

July 1: Site clearing and grading begins at the Shoreham site.

November 8: Lloyd Harbor Study Group issues report strongly criticizing proposed Lloyd Harbor nuclear plant as being "unsafe, unreliable, and unclean." Subsequently LILCO abandons plans for construction of the facility.

December 15: The Advisory Committee on Reactor Safeguards issues a letter favoring the Shoreham construction permit.

1970

September 1: Foundation work for turbine and reactor buildings begins.

September 21: Shoreham's construction permit hearings begin before a United States Atomic Energy Commission Atomic Safety Licensing Board.

1971

May: Atomic Energy Commission begins review of emergency core cooling systems in nuclear plants.

July 23: United States Federal Court of Appeals issues *Calvert Cliffs* decision, requiring that the Atomic Energy Commission prepare environmental impact statements when considering licenses for nuclear plants.

August: AEC regulatory staff issues an analysis saying that the proposed Shoreham emergency core cooling system is acceptably safe.

November 4: Atomic Energy Commission Atomic Safety Licensing Board begins hearings on Shoreham's emergency core cooling system.

1972

June: Staff of Atomic Energy Commission issues Draft Environmental Statement for Shoreham concluding that construction permit is justified.

June: New York State Department of Environmental Conservation concludes that Shoreham can be built without violating State water quality standards.

1973

April 12: Atomic Energy Commission announces it is granting construction permit for Shoreham.

June 14: First concrete is poured for the Shoreham reactor building.

October: First Arab oil embargo begins.

October 26: Atomic Safety and Licensing Appeals Board affirms the issuance of construction permit for Shoreham.

Chapter 4

SHOREHAM'S CONSTRUCTION

The Atomic Energy Commission gave LILCO a construction permit for Shoreham in April, 1973. Over ten years later Shoreham was the most expensive nuclear plant in the nation on the basis of cost per kilowatt (See Table 9, in Chapter 2). A review of the Shoreham project by the New York State Public Service Commission (PSC) concluded in 1985 that about one-third of the plant's costs were incurred imprudently, indicating the scope and magnitude of Shoreham's construction, engineering, and managerial problems. In the 1970s other nuclear projects ran far over budget, and planning and control failures were widespread in the industry (United States Office of Technology Assessment, 1984; Cook, 1985; Leigland and Lamb, 1986; Tomaine, 1987; Campbell, 1988; Morone and Woodhouse, 1989). Thus, while the Shoreham project evolved into a distinctively large failure, many of its fundamental problems were common in the industry.

The Review of Construction

State public utility commissions generally allow utilities to recover construction costs by adding the costs into utilities' rate bases. The utilities are then allowed to charge ratepayers for these costs over the life of the investments. However, utility commissions also can omit costs from a rate base if they conclude that the utilities were wasteful or imprudent in incurring the costs. By 1979 Shoreham's costs and delays had increased to the point that the Public Service Commission undertook such a prudence review.

The review took four years, with court appeals extending to December 1987 (New York State Department of Public Service Staff, 1984 [subsequently cited as Department of Public Service Staff]; New York State Public Service Commission Case 27563, Opinion 85-23, December 16, 1985 [subsequently cited as Public Service Commission]). The Staff of the Department of Public Service (DPS) and others examined internal documents, memoranda, and other materials regarding Shoreham, and interviewed most of the key individuals in the project. The review resulted in almost 12,000 pages of testimony and 2800 exhibits.

The Public Service Commission's standard for evaluating whether the project was prudently managed was established in 1979 in a case involving a 1976 outage at the Indian Point nuclear facility (*Consolidated Edison Company of New York, Inc. Case 27123--Indian Point No. 2 Outage*, 1979). That opinion noted that "[T]he company's conduct should be judged by asking whether the conduct was reasonable at the time, under all the circumstances, considering that the company had to solve its problems prospectively rather than in reliance on hindsight. In effect, our responsibility is to determine how reasonable people would have performed the task that confronted the company" (p. 6).

As a result of the Shoreham review the PSC disallowed $140.8 million in engineering costs, $399.8 million in labor costs, $95 million for repair and replacement of Shoreham's emergency diesel generators, and $759 million in costs associated with the delay of the project (The details of the estimate are presented later in this chapter). The $1.4 billion represented about one-third of Shoreham's costs and, according to the PSC, was a conservative estimate of the project's imprudent management. This chapter examines Shoreham's managerial and regulatory problems during construction.

Oversight by LILCO's Board of Directors and Others

Corporate boards of directors are mandated to review key managerial decisions and, if desirable, modify or redirect them. Shoreham's estimated costs, including the interest on the debt required to finance the plant, rose dramatically throughout the project's life. It was apparent to those involved that the Long Island Lighting Company was controlling the project loosely, at best. Yet, during most of Shoreham's construction, LILCO's Board of Directors failed to recognize effectively the project's devastating impact on the company, let alone redirect it. Relatively passive boards of directors are common in corporate governance; still, it is striking that the LILCO Board missed a problem as large as Shoreham. In turn, LILCO's management did not encourage Board involvement in the project.

Two indicators of Board of Directors involvement are the amount of information provided to the Directors prior to meetings and the percentage of the meetings devoted to discussion of Shoreham. "Board Books" contain the written information provided to each member prior to Board meetings; the minutes of the meetings themselves indicate the percentage of discussion at the meetings devoted to each topic. Table 10 summarizes both indicators from 1965 to 1983. Table 10 also indicates the contemporary estimates of Shoreham completion costs, and Shoreham's

Table 10. LILCO Board of Directors Activity and Shoreham Costs, 1965-1983

	% of Board Information on Shoreham	% of Board Activity on Shoreham	Shoreham $ Estimate (thousands)	% of Total Plant
1965	--%	-%	$70,000	--%
1967	2	7	--	--
1968	2	5	170,000	--
1969	2	5	261,000	32
1970	11	13	300,000	34
1971	7	5	350,000	38
1972	6	7	350,000	34
1973	5	11	506,000	48
1974	13	14	695,000	62
1975	9	3	695,000	60
1976	13	4	969,000	81
1977	8	1	1,223,000	91
1978	5	3	1,337,000	93
1979	9	4	1,581,000	106
1980	11	5	2,213,000	142
1981	10	10	2,490,000	156
1982	20	16	3,080,000	188
1983	14	15	3,400,000	205

Source: Department of Public Service Staff, 1984: Tables One and Two (pp. 52, 55).

estimated costs *compared to the value of LILCO's total plant assets in each year.*

The information LILCO management provided to the Board members prior to the meetings, and the discussions at the meetings themselves, did not focus substantially on Shoreham even when the plant came to dominate LILCO's financial situation. Thus, this level of "external" review was not functioning effectively. The Department of Public Service Staff noted that "Putting it bluntly, at some time in the history of Shoreham, the Board of Directors should have demanded a full examination of the project by independent auditors and consultants. At a minimum, the Board should have rigorously investigated the qualifications and performance of LILCO's project management and senior management teams, the routine failure to meet cost and schedule objectives, and the estimating and scheduling techniques" (1984: 58).

Public utility commissions evaluate requests for rate increases made by electric utilities, and so the New York State Public Service Commission itself had the authority to review LILCO's management practices. In the late 1970s the PSC began to scrutinize LILCO's performance when it undertook the prudence review.

The PSC had granted LILCO rate increases through the 1970s (See Table 11). The PSC also had allowed LILCO to recover part of Shoreham's costs through rates even though the plant was neither completed nor operating. Normally for a utility to receive a return on an investment in a power plant the plant must be "used and useful." However, nuclear projects take far longer to complete and are vastly more expensive to build than fossil fuel plants. If a utility could receive no cash return on a nuclear power plant until it was operating the utility would have a large share of its capital tied up in an asset producing no cash revenue. The Public Service Commission tried to mitigate this effect by allowing LILCO to include a portion of Shoreham's cost in its rate base (This is called allowing "construction work in progress," or CWIP, into the rate base). In 1976 the PSC allowed $55 million of Shoreham's costs into its rate base, and allowed another $200 million of CWIP in 1979.

But by the end of the decade the PSC was looking critically at LILCO's escalating rate requests. It approved only about one-sixth of a rate request in 1979, and that year also rejected LILCO's request to include an additional $400 million of CWIP in its rate base. Writing of the nuclear industry in general, James Cook notes that "Regulators saw what was coming. As costs began to rise into the billions and the utilities began arguing for more generous CWIP, the state commissions became increasingly unsympathetic" (1985: 94). On March 6, 1979 the State Department of Public Service (DPS) called for the prudence review--later approved--"to determine to what extent mismanagement by the Long Island Lighting Company might have led to cost overruns of nearly 500 percent" at Shoreham. LILCO argued that its cost overruns were "about average" and due to inflation, regulatory changes, and environmental hearings, but in calling for the review the DPS staff cited internal LILCO studies suggesting that "factors such as worker productivity that were under the control of Lilco may have contributed to both the delays and cost increases" (*New York Times*, March 7, 1979: B1).

As noted above, the review took four years, with court appeals through 1987. The Public Service Commission subsequently forced LILCO to

Table 11. Rate Increases Granted to LILCO By Public Service Commission

	Rate % Increase		
1970	5.8%	1980	3.2%
1971	.9	1981	14.4
1972	2.3	1982	.3
1973	5.5	1983	1.1
1974	0	1984	16.2
1975	20.9	1985	.0
1976	5.8	1986	4.0
1977	1.1	1987	.0
1978	8.3	1988	.0
1979	3.3		

Source: New York State Public Service Commission, reprinted in *Newsday*, May 27, 1988.

absorb about one-third of Shoreham's costs, and the review documented the need for major management improvements at LILCO.

LILCO's Management of the Regulatory System

Federal regulation fundamentally shapes the construction of nuclear power plants. The Atomic Energy Commission, and later the Nuclear Regulatory Commission (NRC), determined how a nuclear plant would be designed and operated as much as the owners and builders did. As discussed in Chapter 2, LILCO's Arthur Sugden had recommmended in 1968 that the Shoreham project be deferred partly because of uncertainties about Federal regulation.

The increase in Federal regulatory actions in the late 1960's-early 1970's confirmed Sugden's concerns about regulatory changes. For example, the number of Regulatory Guides issued by the Atomic Energy Commission increased from three in 1970 to 21 in 1971 and 33 in 1972 (Komanoff, 1981: 51). The New York State Department of Public Service Staff investigation of Shoreham noted that "While many of these new regulations and guides surfaced in the early 1970's, they were under study as proposals in various industry code committees well before they were adopted by the AEC. Stone and Webster, as an architect/engineer, sat on many of these committees and was directly aware of such developments in the nuclear industry" (1984: 146). Regulatory changes inevitably would affect the project's cost and duration. However, the impact could have been reduced if Stone and Webster and LILCO had identified the likely changes and, in a proactive way, integrated them quickly into the project's engineering. Instead, the companies in some cases tried unsuccessfully to shape how the NRC applied regulations to Shoreham and, in other cases, did not deal with the changes sufficiently quickly. Construction times and costs accordingly increased. Examples of the problems are discussed below.

New Piping Standards

After about three years of deliberations, the Atomic Energy Commission changed the requirements for piping systems in nuclear power plants in June 1971. Shoreham's original design used the American National Standards Institute (ANSI) Piping Code numbered B31.1. In February, 1968 ANSI proposed a new Piping Code numbered B31.7, and requested comments on the proposal. The American Society of Mechanical Engineers (ASME) Nuclear Pump and Valve Code, referred to as the ASME III code, included the ANSI B31.7 Piping Code.

In May 1969, General Electric, the manufacturer of Shoreham's Mark II system, told the Stone and Webster project engineer that the Atomic Energy Commission had requested that nuclear plants comply with the ASME III code, and thus the new B31.7 Piping Code. One month later the AEC told Stone and Webster and LILCO that the new piping code might be required for Shoreham. In November 1969, the AEC proposed formally that all nuclear power plants receiving construction permits after April 1, 1970 comply with the ASME III standard, which included Piping Code B31.7. In June 1971 the AEC adopted a final regulation specifying that all nuclear plants receiving construction permits after January 1, 1971 had to comply with the new piping code.

The new regulations for piping would require major changes in the engineering plans for the plant. However, as discussed later, for various reasons LILCO temporarily suspended engineering work at Shoreham in August 1971, and at that point the related changes had not been made in Shoreham's plans. In fact, when engineering resumed in 1972 LILCO discovered that the percentage of the engineering plans actually completed was far lower than had been thought prior to the shutdown. (This point also is discussed below.)

The Staff of the Department of Public Service (DPS) maintained that

> Instead of ordering Stone and Webster to begin the process of incorporating ASME III into Shoreham's design once the NRC announced the proposed regulation, the company procrastinated, and by the time engineering was shutdown in August 1971, the new code had not been incorporated into Shoreham's design...LILCO waited much longer than it should have to implement Piping Code B31.7. The signals were clear from the American National Standards Institute in 1968 and the NRC (sic) in 1969 that this code would be the design basis for nuclear plants. Instead of analyzing the impact of this new code on the Shoreham project, LILCO took an unstudied risk that it would not be subject to this new code (Department of Public Service Staff, 1984: 166-167).

LILCO responded that by September 1970 Stone and Webster in fact had reclassified piping systems to accomodate the new standard. But, if that were the case, engineering faced a large task in actually incorporating the new standard for piping systems into the plant's design and construction, and LILCO should not have suspended engineering in August 1971. The DPS Staff noted "This work [on the piping system] represented a major engineering effort, and was one of the main reasons that S&W's estimate of the percentage of engineering complete was inaccurate...This major miscalculation is part of the reason that engineering failed to support field construction once the project was restarted in 1972...there is no indication that the extensive work required to comply with ASME III was taken into account when the decision to shut down was made" (1984: 167).

Pipe Break Outside Containment

In December 1972, the Atomic Energy Commission required LILCO and other utilities to analyze the effects of a possible accident involving breaks in a high energy piping system. The utilities had to determine the damage that might result from the whipping action of broken pipes, from the discharge of pressurized steam or water, and from changes in pressure, and how to prevent the damage. This might require changes in the design and layout of a nuclear plant, and was called the "pipe break outside containment," or PBOC, issue. The Atomic Energy Commission told LILCO that any necessary redesign of piping should follow new AEC criteria for preventing damage from such pipe breaks.

Following the release of the AEC's Pipe Break Outside Containment criteria in 1972, Stone and Webster prepared a procedure for analyzing pipe breaks that might apply to several plants on which it was working. LILCO submitted this "generic" analysis of PBOC in nuclear plants to the Atomic Energy Commission as part of its Shoreham operating license application. The AEC told LILCO that it would study the analysis when it reviewed Shoreham's Final Safety Analysis Report.

As long as the AEC (later the NRC) had not formally approved the pipe break analysis, there was a risk that the AEC would require major redesign of the piping system. This was a substantial risk, because LILCO had based its submission on Stone and Webster's *generic* analysis of PBOC requirements, rather than a Shoreham-specific analysis as requested by the AEC. The DPS Staff argued that "it was wrong for LILCO's management not to press the NRC for an early resolution of Shoreham's PBOC requirements soon after Stone and Webster completed its report in February 1974. LILCO exposed the Shoreham project to a substantial risk of backfitting if the NRC did not agree with Stone and Webster's generic interpretation" (1984: 169).

In 1975 the NRC rejected LILCO's Final Safety Analysis Report (FSAR) partly because of what it called inadequate information on the plans to handle Pipe Break Outside Containment at Shoreham. The NRC asked LILCO for additional information. LILCO denied that the information was inadequate; the NRC rejected the denial. LILCO questioned the NRC's view again, and said that the plant had been designed and was being built using an acceptable interpretation of piping standards. In 1976 the NRC again rejected the Stone and Webster/LILCO analysis for the Shoreham plant. LILCO finally accepted the NRC's standards for analysing Pipe Break Outside Containment, and submitted its revised study in December 1978. By the time the issue was settled the NRC required LILCO to redesign and replace a substantial amount of piping-related equipment.

LILCO maintained that the Nuclear Regulatory Commission had explicitly concurred with its original PBOC analysis in meetings, and implicitly approved it by approving other projects using Stone and Webster's generic analysis. LILCO also pointed out that General Electric, the manufacturer of the Mark II system, had approved the analysis. However, the Department of Public Service Staff reviewed notes and memoranda from the related meetings and concluded that they--and NRC comments--did not show any such NRC approval of the Shoreham PBOC analysis. The DPS Staff also said that General Electric's approval of the analysis was not relevant because General Electric was not the regulator which ultimately had to grant a license to the plant (pp. 169-173).

The Pressure Suppression Issue

As discussed in Chapter 2, General Electric had designed the Mark I, Mark

II, and Mark III containment systems. The original proposal for Shoreham was made with the Mark I design, but the Mark II system was actually used in the plant. Each of these systems used a pool of water to absorb unusual increases in steam pressure.

In 1975 General Electric built a facility to conduct tests on the new Mark III design. The tests showed that in a loss of coolant accident (LOCA) there could be an unexpectedly large rush of air and steam into the pressure suppression pool. The steam could condense in the pool more violently than previously believed, and the resulting pressures could shake the entire reactor building and possibly damage critical equipment.

The Mark I and Mark II designs shared this problem with the Mark III. The NRC subsequently required that all of the related containment structures had to be strengthened to reduce the chance of equipment failures. Steel was added to reinforce concrete containment walls, steel containment walls were thickened, and supports for equipment around the pressure suppression pool were added (Komanoff, 1981: 53). In 1975, LILCO formed a group with other Mark II owners to design generic solutions to the Mark II pressure suppression problems. In 1978, the NRC indicated that it would be willing to quickly address these problems for plants expecting operating licenses in the near future, and at the time Shoreham was one such plant.

However, LILCO did not get the NRC's approval of its methods for measuring the hydrodynamic loads at the Shoreham plant. Disagreements over how to measure Shoreham's hydrodynamic loads kept the NRC from approving LILCO's solution to the pressure suppression problem until 1982. Eventually, additional pipes to diffuse steam discharges into the suppression pool were added to the plant, pipe layouts were changed, and about 25,000 seismic restraints were added throughout the plant to prevent damage from both hydrodynamic loads and earthquakes. The Public Service Commission maintained that LILCO's resistance of the NRC staff on the Mark II question was one of the major causes of Shoreham's costly delays in the early 1980's. In contrast, the PSC pointed out, Pennsylvania Power and Light's Susquehanna nuclear project obtained a construction permit seven months after Shoreham's permit was issued, dealt efficiently with the same Mark II problem, and entered commercial operation in June 1983 (Public Service Commission, 1985: 54-57; Fitzgerald, 1987: 31).

The Size of the Reactor Building

As discussed in Chapter 2, LILCO originally planned to build a 540 megawatt station at Shoreham. The reactor building, which would contain the reactor, the primary containment structure, and related equipment, was meant to accomodate a 540 MW system. When LILCO decided in 1968 to upgrade the generating capacity to 820 megawatts it decided not to change the plans for the diameter of the reactor building; Stone and Webster had advised LILCO that the 820 MW system could be accomodated into the original building. However, the upgrading left the Shoreham building unusually small. A number of Mark II plants being designed at the time had buildings fifty percent larger than Shoreham's. The Nine Mile II plant in Oswego, New York, also a Stone and Webster project, had an area inside the containment structure eighty percent larger than Shoreham's. The DPS Staff presented evidence that LILCO performed no calculations showing how the upsizing would affect the construction effort (Department of Public Service Staff, 1984: 194).

The relatively small size of the reactor building aggravated LILCO's problems in managing regulatory changes. Most of the changes required the addition of equipment, piping, restraints, and other structures. The amount of space inside the building severely constrained the routing of pipes, cables, and the location of equipment, and also limited the ability of workers to maneuver. Consequently, the size of the reactor building made the changes more onerous than they otherwise would have been. This, combined with LILCO's failure to adopt the ASME III piping regulations on a timely basis, proved to be a major problem for construction and engineering.

The decision not to increase the size of the reactor building following the power upgrade was disturbing particularly because LILCO and Stone and Webster were aware, even before concrete was poured for the building, that the Atomic Energy Commission was about to make significant regulatory changes. A number of these changes were made between 1969 and mid-1973, and the building's foundation was not actually set until June 1973.

LILCO's Relations With the NRC

The Department of Public Service Staff reviewed other instances in which LILCO failed to comply with the NRC's regulations, resisted NRC requests for modifications, and then ultimately complied after substantially reworking parts of the plant (For example, for cases involving separation of electrical cables and conduits, cable tray supports, and the handling of exceptions from NRC standards [1984: 174-180]). Generally, LILCO's relationship with the NRC during construction was not good.

Following the Three Mile Island Accident the NRC began to issue periodic performance evaluations of utilities building nuclear power plants. The reports were called "Systematic Assessments of Licensing Performance," or SALPs. From 1980 through 1983 the reports suggested that the NRC regarded LILCO as being "an active and technically knowledgeable applicant, however, they lack [boiling water reactor] operating experience and they are frequently recalcitrant" (United States Nuclear Regulatory Commission, 1981b: 2; subsequently cited as Nuclear Regulatory Commission).

The NRC's project manager commented on the first SALP regarding the difficulty of getting information on the project from LILCO, and said that "This applicant opposes most Staff requirements and tries to get by with the minimum to resolve staff concerns...Strictly reactive, they will not do it unless you spell it out for them" (Nuclear Regulatory Commission, 1981a: 1). He also said that "The applicant had many long-standing open items throughout this appraisal period. Because the applicant had not neared the completion of construction, they opposed many staff positions in the hope that the staff would back off." The second SALP noted that "These results continue to show delays and in some cases inadequate resolution of NRC identified problems as well as an apparent lack of aggressiveness by the licensee in obtaining design conformance to the [Final Safety Analysis Report]. The above items are also indicative of the ineffective and delayed corrective action by S&W engineering" (Nuclear Regulatory Commission, 1982: 16). The SALP Project Manager also noted problems in obtaining information from LILCO and the company's "frequently recalcitrant" attitude in meetings (1981b).

According to a Stone and Webster employee's notes for a meeting with the NRC to discuss the third SALP Report, the NRC's Chief of Reactor Licensing said that "The NRC is not pleased that it is required to spend so much time in reviews at the Shoreham plant. The NRC believes that LILCO should do much more to insure that the work is done properly and that items are ready when presented to the NRC for inspection...The attitude of LILCO's management needs to be improved." The NRC's Deputy Director of Licensing was noted as saying "although he was not a member of the SALP committee, in reading the report he is convinced that many of the Category 2 ratings [average] received by LILCO were really on the borderline between Category 2 and 3 [below average]. It was his opinion that the SALP committee had given LILCO the benefit of the doubt for most of the Category 2 ratings...He stated that LILCO must become pro-active in insuring that it meets requirements as opposed to its current trend of being reactive to NRC comments and concerns" (E.J. Brabazon memo to P.A. Wild, June 3, 1983).

The Failure of Engineering to Support Construction

Building the plant efficiently required that one knew reasonably well how the plant's components were to be designed and integrated before construction. The DPS Staff report in 1984 observed that "Nuclear power plant construction requires continual interaction by engineering with procurement, licensing, and construction groups. Engineering must effectively accomodate the demands of regulators and the needs of construction to assure that the engineering and design of the plant is completed in a timely fashion. As it is the central cog of nuclear construction, the effectiveness of engineering management is a prime factor in determining how long it will take to complete construction and at what cost" (1984: 188).

Engineering failed to support construction schedule through most of the Shoreham project. LILCO's problems in managing regulation contributed to the engineering situation. Also contributing to the engineering lag was the fact that LILCO suspended engineering preparations between September 1971 and September 1972, setting engineering back at a time when it was already seriously incomplete. Consequently, either the construction was delayed, or was completed but then in many cases had to be reworked.

The Decision to Suspend Engineering

On August 3, 1971 LILCO told Stone and Webster to halt construction, engineering, and procurement for Shoreham. Stone and Webster assigned its Shoreham work group to other activities, leaving a small group for recordkeeping and support of the licensing application. LILCO authorized Stone and Webster to resume engineering in May of 1972, and construction resumed on October 1.

LILCO said that it suspended engineering in August 1971 because of the U.S. Court of Appeals decision on *Calvert Cliffs* in July (449 F. 2d 1109 [D.C. Cir., 1971]). The *Calvert Cliffs* decision required the Atomic Energy Commission to fully comply with the National Environmental Policy Act (NEPA) when licensing the construction or operation of nuclear power plants. Accordingly, the AEC would have to prepare an environmental impact statement for Shoreham's construction permit hearings, further extending what was already the longest construction permit proceeding.

Shoreham's construction permit hearings had started in September 1970. They were expected to last about one month, but by the Calvert Cliffs decision had taken ten months. The NEPA requirement was now another hurdle confronting Shoreham. Explaining the engineering suspension, LILCO argued in 1984 that by August 1971 it feared that Shoreham might never receive a construction permit. Thus, the company stopped ordering equipment that might have to be abandoned if the plant could not be built; the suspension of equipment orders in turn stopped the flow of information needed for engineering (Department of Public Service Staff, 1984: 198). Furthermore, the company said that it believed the shutdown would not affect the project very much, even if a construction permit was issued, because at the time it believed that engineering was advanced enough to support construction. At shutdown LILCO estimated that Shoreham's engineering was 55 percent complete overall, and that the engineering for the structure of the plant was more advanced than this.

The DPS Staff rejected both of LILCO's claims about its reasons for the engineering suspension. First, the Staff argued that LILCO never really expected that the *Calvert Cliffs* decision would block Shoreham's construction. After examining LILCO's documents the report stated that "The contemporaneous documents show LILCO management was concerned over the *delay* in receiving a CP for Shoreham, but there is no evidence of any serious thought or planning for the possibility that a CP would not be issued. The company's contention is a rationalization rather than a true representation of management's thinking at the time" (1984: 210).

Second, the DPS Staff argued that Shoreham's engineering was already running behind schedule prior to the suspension. In July 1971, LILCO said that it hoped to complete eighty percent of the engineering for the plant before Shoreham actually received a construction permit. Stone and Webster told LILCO--and LILCO claimed to the Department of Public Service--that engineering was 55 percent complete at the time of the engineering shutdown, and that the structural engineering for the plant was more advanced than this. But the 55 percent estimate of completed engineering was too high. Stone and Webster's estimates were based only on an estimate of man hours used in engineering, rather than a direct examination of the status of actual engineering drawings. Stone and Webster did provide LILCO a Summary of Drawing Progress table indicating the percentage of drawings issued compared to the number of drawings scheduled. But the Summaries "give no indication of whether issued drawings are actually complete...Had LILCO engineers simply taken the time to review the pertinent Drawing Schedule and Progress charts they would have seen that a significant portion of the concrete and structural drawings issued at the time of the shutdown were not complete and required revision before they could be used by construction...The Level III charts show that most of the work for concrete drawings had not been done. Thus, reasonable management would have known at the time of the shutdown that the estimate of 55% of engineering complete was highly inaccurate" (pp. 202-203).

Also, much of the engineering that had been done prior to the suspension had not incorporated regulatory changes in place or that would probably be adopted, such as the requirements of ASME III for piping and mechanical systems (adopted in June, 1971): "[T]he 55% engineering complete estimates [that LILCO] was receiving from Stone and Webster did not reflect the impact of ASME III on engineering already performed. The company, therefore, should have realized that Stone and Webster's engineering complete estimates, and particularly the mechanical and structural estimates, were vastly overstated. This alone should have mandated a continuation of engineering" (Department of Public Service Staff, 1984: 204).

The Dispersion of the Stone and Webster Site Team

When LILCO told Stone and Webster to suspend engineering, Stone and Webster reassigned most of its experienced Shoreham staff to other projects. When the project resumed, a staff of similar quality could not be assembled.

In 1971 Stone and Webster was working on the Beaver Valley I, North Anna Units 1 and 2, Fitzpatrick, and Shoreham facilities. In September, 1971, a month before the actual shutdown, Stone and Webster became the architect/engineer (A/E) for the Beaver Valley nuclear project. In October 1971 it announced that it would be A/E for Nine Mile 2 in Oswego, N.Y., and in December 1971 announced that it would be A/E for River Bend Units I and 2. When LILCO suspended engineering Stone and Webster reassigned its Shoreham staff to other projects.

Given these commitments to other nuclear projects there was no guarantee that Stone and Webster would be able to return an experienced staff to Shoreham when the project resumed fully. LILCO neither asked for nor received assurances of the quality of staff that would return to Shoreham when the project resumed. When the project resumed in 1972 one LILCO official complained that "S&W appears to be experiencing some difficulty in putting a design group together" (A.J. Wofford memo to A.T. Jorgenson October 26, 1972: 2). Another said that neither LILCO nor Stone and Webster provided sufficient staff to reactivate the project. An internal LILCO report from its Nuclear Projects Department in May 1973, prepared one month after Shoreham received its construction permit, said that "The construction effort is being impeded by engineering and after all the excuses are made and accepted the fact remains. Stone and Webster does not appear to be adequately staffed to support a tight construction schedule. S&W intends to assign additional structural designers and a full time Project Engineer to resolve this problem" (Nuclear Projects Department Report to LILCO management, May 21, 1973: 6). Two months later the Nuclear Projects Department reported that "S&W does not have an engineering staff with the depth and experience needed to concurrently resolve the day-to-day construction problems while maintaining their long-range schedule...Engineering progress is not sufficiently ahead of construction to permit an orderly, well-planned building program. There is every reason to believe that engineering holds will continue to hamper the construction efforts during the next few years. This situation may, in a large measure, be due to the delays and shutdowns which enabled new codes and standards to be implemented and which left us with few of the talented and experienced engineers who began this project" (July 23, 1973: 3-4). LILCO's manager after the reactivitation, Mr. T.W. Burke, said that "The remobilized Stone and Webster project was staffed with less experienced engineers...The less experienced staff further compounded our problems by being slow to recognize the impact of the AEC inspired changes in pipe break and seismic requirements, and once these impacts were identified the inexperienced engineers were unable to speedily cope with the ramifications" (memo from Burke to A.W. Wofford, March 19, 1974, p. 5).

The Effects of the Engineering Lag: Two Examples

Metal Embedments

The problems of installing metal embedments at Shoreham illustrate the effects of the mismatch between engineering and construction progress. Embedments are metal plates which are attached to rebar (the steel frame) prior to concrete being poured. Once concrete is poured, the embedments become a part of the wall of the reactor building. Builders attach pipe hangers, pipe restraints, or other apparatus on metal embedments.

One of the factors determining the number and placements of embedments was naturally the number, location, and type of pipe hangers and pipe restraints. However, Stone and Webster's engineers had not completed the analysis of pipe stress or restraint requirements prior to the pouring of concrete, and so they "tried to guess the numbers of locations of embedments" (Department of Public Service Staff, 1984: 228). The construction drawings were completed despite the incomplete analysis of embedment requirements; "Thus, substantial numbers of embedments simply were not included in the initial design because the engineering effort was behind and management was committed to trying to maintain the structural construction goals and pressure vessel set. Dr. Cordaro [of LILCO] acknowledged that only 10% of the embedments ultimately required for the project were provided in the initial design" (p. 229).

Courter and Company, the piping contractor for Shoreham, pointed out that the lack of embedment plates meant that it had to design metal base plates for pipe hangers, and drill holes in the wall to attach fasteners. Furthermore, the drilling might cut the metal rebar, possibly weakening the structure, and so "it was almost impossible to install a base plate without the need to drill additional holes to avoid cutting rebar. In every instance, the job stopped until an engineering analysis could be made to determine whether or not the rebar could be cut. If embedment plates were available, all that would have been required would have been simple welding of the structural members to the embedment plate, eliminating all of the fabrication, drilling, and engineering delays associated with concrete fasteners" (memo from Courter and Company to LILCO on Job Cost Overrun, April 21, 1981: 1).

As these sorts of problems surfaced, engineering drawings were hurriedly issued, often prematurely, to try to cope with the lag. The engineering then would have to be further revised. LILCO's Nuclear Project's Department reported on April 26, 1974 that "The major engineering problems at this time relate to the fact that engineering is just days and in some cases only hours ahead of the construction effort" (p. 7 of April 26, 1974 report of the Department to LILCO management).

Field-Generated Engineering Changes

Engineering changes by Stone and Webster, LILCO, or contractors were made through Engineering and Design Coordination Reports (E&DCRs). Stone and Webster, as overall project manager, had to approve the changes. Approximately 70,000 E&CDRs were generated at Shoreham between December 1979 and March 1983; that figure excluded changes made prior to and after that period. The Department of Public Service Staff maintained that "The number of design changes was unreasonably high because a significant portion of these changes were due to errors or noncompatibilities (inconsistencies between design documents) that management should have, but did not control...Shoreham's engineering was hopelessly behind at the construction start. This meant that engineering was constantly in the position of playing 'catch up' in order to support construction. And, as a consequence, the engineering supplied to the field was typically late and/or deficient" (pp. 246-247).

A crude comparison concluded that the number of design changes at Shoreham exceeded the number at other nuclear power plants by a factor of two or three. Also, at Shoreham *personnel in the field* rather than the architect engineer's *central office* were far more likely to initiate the changes. The ratio of field-generated to A/E initiated changes was almost six to one, compared to 1.5 to one at other plants. The large ratio of field-generated to A/E generated changes was important because it suggested that errors rather than regulatory changes were the largest source of change. Stone and Webster's Boston office rather than field engineers handled regulatory changes, so the very large number of field changes arguably reflected "inefficient engineering" (Public Service Commission, 1985: 53). A study by Stone and Webster in 1975 noted that 45 percent of a sample of E & DCR's "were required due to errors or non-compatibilities in released documents" (letter from Cusick to C.E. Goodman of Stone and Webster, January 8, 1976). A study about one year later showed similar results.

LILCO's Relationship with Stone and Webster

LILCO decided to hire Stone and Webster as a construction manager because LILCO lacked the experience and resources necessary to build a nuclear power plant. Still, the company planned to rely on nuclear power more in the future, and so wanted to develop its own nuclear construction experience. Although under the contract Stone and Webster would be the project's construction manager, LILCO wanted to be involved closely to acquire this experience.

However, one of the problems over the life of the project was a serious conflict over precisely who -- LILCO, Stone and Webster, or subcontractors -- had authority for particular issues as they emerged. As early as 1970 Stone and Webster officials complained that the company had to "feel [its] way in a manner satisfactory to LILCO and be responsive to many directions being given from various sources...It was unclear at this time which functions of engineering, construction, and purchasing would be the responsibility of LILCO or S&W" (Stone and Webster notes on S&W-LILCO conference on Shoreham, February 16, 1970: 3-4).

Although Stone and Webster may not have felt in charge, neither did LILCO. LILCO tried to monitor Stone and Webster's activities, but Stone and Webster did not go out of its way to cooperate with LILCO's oversight. LILCO, in turn, assigned few staff to monitor the project. In 1973 LILCO had assigned only three monitors to the Shoreham site. In March, 1974, the monitoring group reported that it could not obtain complete information on Stone and Webster's field policies for construction; that little information on costs had been gathered; that "no significant input has been reported by the construction group, due to the secrecy of the constructor;" and that there was a "Lack of LILCO personnel to follow work by discipline" and a state of "Constructor;s [sic] secrecy and poor reporting" (LILCO Nuclear Projects Construction Group Self Analysis Report, March 26, 1974: I-1-4). By December 1975 LILCO increased the monitoring staff to 13. By that time heavy construction had been underway for two and one-half years, and the Department of Public Service Staff argued that it was unreasonable for LILCO to have waited so long to increase its monitoring staff on site (p. 123).

In 1975 LILCO's project manager also began allowing individual project contractors to bring their problems to him rather than to Stone and Webster, undercutting what control Stone and Webster did have and increasing the antagonism between the two firms. Even LILCO officials acknowledged that the situation created "confusion over who was in charge," and a situation as one in which "some troops didn't know who the leader was" (cited in Department of Public Service Staff, 1984: 83).

In May 1976 LILCO partially reorganized the project, creating the "Unified Constructor Organization" (UNICO). Under UNICO, LILCO's current monitors would move away from a monitoring role and directly into middle-level line responsibilities in Stone and Webster's site construction operations. LILCO and Stone and Webster both acknowledged at the time that they had in effect been competing in the construction manager role, and the integration of LILCO's staff into Stone and Webster's site organization presumably would ease the conflict.

However, the UNICO actually reduced LILCO's control over the project. First, LILCO's thirteen person monitoring staff was taken out of its broad oversight role and put into narrow line responsibilities. LILCO's project manager evaluated the UNICO experience by noting that the reorganization had diminished the level of hostility on site between the two companies, and that "In my view UNICO represents a significant improvement over our previous organization." Nevertheless,

> [t]he major disadvantage of UNICO is the fact that we have sharply diminished our ability to identify and analyze construction problems from an owner's point of view. I had hoped that a combination of the various management information systems and my direct dealings with the Stone and Webster Resident Manager [the senior S&W site construction manager] would provide adequate information to properly manage the overall project. Although my communications with the Resident Manager have improved considerably he still tells me what he wants me to hear. The elimination of our construction criticism from the LILCO construction group represents a net loss in our ability to identify and resolve problems. By the time we become aware of a serious problem, such as the reduced piping installation rate during July and August, it is necessary to form a task force on a crash basis and to ferret out the causes and recommend solution" (memo from T.J. Burke to A.W. Wofford, August 31, 1976: 3).

Second, the movement of LILCO's monitoring staff to line responsibilities produced a net loss of control because the line positions were not sufficiently high to allow LILCO's staff to exert much influence. LILCO's Mechanical Construction Engineer commented that "It is my opinion that since LILCO personnel will not be placed in any position which can directly affect overall job policies and decisions, the possibility of our improving the current outlook for the construction effort is minimal" (memo from W.J. Museler to T.J. Burke, April 23, 1976). The memo from Burke to Wofford, referenced above, commented that "I would agree with those interviewed that the infusion of LILCO personnel has had little overall impact on the

S&W Constructor Management team. The Stone and Webster Constructor engineers are already rather well motivated and a few LILCO engineers would not be enough to make a noticeable impact on the group."

Even after the establishment of UNICO, the project still was characterized by a confusion of responsibilities, antagonism among LILCO, Stone and Webster, and key contractors, and a lack of clear direction and leadership (Department of Public Service Staff, 1984: 91-103). A report by Arthur D. Little to LILCO and Stone and Webster in 1977 concluded that

> There are still, in fact, two project managements at work at Shoreham. Neither is in charge in the sense of having clear-cut responsibility and authority to take actions necessary to get the job finished.
>
> The dual managements are not in accord with the decision which we understand was reached at the January 6, 1977 meeting of senior executives of the companies: that S&W is to be the exclusive manager of the project. The perceptions, observed behavior, and responses of LILCO, S&W, and the contractor personnel at the workshop confirm that S&W is not effectively in charge.
>
> One reason S&W has not assumed full control is the unclear line of authority over the mechanical and electrical contractors. We understand several approaches on this problem are being considered: we urge LILCO and S&W to resolve the problem quickly.
>
> However, we also caution against what we sense to be a growing overoptimism that resolution of the contracting relationship will "cure" Shoreham's problems. They will not be cured until there is a more fundamental, top-level delineation by S&W and LILCO of the roles, responsibilities, and authorities of the two organizations, and until these determinations have been unambiguously communicated to all levels of management and supervision involved in this project, including contractors (Arthur D. Little, Inc., 1977: 2-3).

Labor Productivity

Labor productivity at the Shoreham site generally was acknowledged to be extraordinarily low. The internal and external reviews of the site present a situation in which large numbers of workers arrived late, quit early, took very long breaks and lunches, commonly stole materials from the site, and frequently used alcohol or drugs on site. There was also widespread sabatoge--such as scarring of welds--to prolong the life of the project (Diamond, 1981: 21R). The problems were noted beginning in 1974, and observed throughout Shoreham's construction.

Welding productivity in 1976, for example, was only one half of that required to maintain the schedule (S&W notes of conference on Shoreham, July 13, 1976: 2). A memo from LILCO's resident manager said that "An intolerable amount of work time is being lost due to late starts, early quits, and extended coffee breaks. Assuming that this lost work time averages an hour per day per man--and it may well be more than that, if the truth were known--that's 2,200 manhours per day that are lost to this project, and over $35,000 cost per day to LILCO or to lump sum contractors, for which no value is received" (memo from R.L. Cusick to site personnel, August 23, 1976). The Department of Public Service added that, assuming 200 workdays per year, such a loss would be 4.4 million lost craft manhours over the ten year life of the project.

In 1977, LILCO conducted work force sampling studies to measure productivity. The studies suggested that craftsworkers were engaged in direct work only 21 percent of the time; that late starts and early quits accounted for 8.1 percent of crafts hours; and that workers were "waiting around" about 41 percent of the workday. In contrast, direct work at other large construction sites averaged 31.5 percent, and direct work percentages of 40-44 percent had been attained at some other nuclear construction sites (UNICO Shoreham Work Sampling Study, November 14-December, 1977 [undated], p. 8). In 1978 LILCO reported that 112 welders welded five days or less in February, and that 35 did no welding at all; Courter and Company, the piping contractor, reported reject rates of 94 percent for field welds in the reactor area (Department of Public Service Staff, 1984: 289-293). Also in 1978, LILCO estimated that by the time Shoreham was completed it would require 39.5 person-hours per each kilowatt of generating capacity; the average for plants beginning in operation in 1981 was 12-15 person-hours per kilowatt, and for plants expected to begin operation in 1990 was 18-21 man-hours per kilowatt (Diamond, 1981: 20R).

Low labor productivity was almost inevitable in a situation as chaotic as Shoreham's. As discussed earlier, The relations between LILCO, Stone and Webster, and the NRC were dismal, aggravating the effects of regulatory changes. Engineering was too far behind to support construction, and in the attempts to catch up the hurried work was flawed and commonly needed to be redone. LILCO, Stone and Webster, and subcontractors clashed over their responsibilities and authority. There was no effective leadership at the highly complex project which might have brought some order to the situation. In this environment of regulatory turbulence, engineering problems, inter-firm hostility and conflicts over authority it was not surprising that productivity was low. Labor leaders could argue that they were not delivered work or equipment in an orderly fashion. Virtually everyone agreed that there was minimal supervision of workers on the job.

The Failure of the Diesel Generators

The NRC's regulations required a nuclear power plant to have a back-up power source which could operate safety systems in case primary power systems failed. To meet this regulation LILCO purchased three emergency diesel generators from Transamerica Delaval, Inc (Delaval). The diesels were delivered to the site in 1976. In August 1983, after the construction phase of Shoreham was essentially complete, the crankshaft in one of the three engines broke during operational testing. Inspections of the other two generators revealed that their crankshafts also had cracked. An investigation revealed that the crankshafts were undersized, and so could not stand the stresses of normal operation. The investigation uncovered other damage as well. LILCO repaired the diesels, but subsequently purchased replacements from Colt Industries. The breakdown of the emergency diesel generators delayed testing of the plant by about sixteen months. Repairing the diesels, and purchasing and building facilities for the replacements, added about $95 million to the cost of the plant. The costs of the related delay ran into the hundreds of millions of dollars. Breakdown of the emergency diesel generators was one of the most costly events in the latter stages of the Shoreham project in terms of money, time, and political credibility. The costs in money and time were clear. Also, LILCO, which had been vigorously defending the safety of the plant, was left to explain why the emergency back-up power system--which represented a rather simple technology--had failed during testing.

From initial purchase through testing the diesel engines were suspect. The Department of Public Service Staff criticized the handling of the diesels on two central grounds. First, the DPS staff argued that the initial procurement of the diesels from Delaval was imprudent. LILCO selected Delaval's bid over the bids of Colt Industries and Cooper-Bessemer. Delaval's initial bid was only 4.8 percent less than Colt Industries, and 2.5 percent less than Cooper-Bessemer's. Since these were only initial bids and were subject to negotiation and change, giving the order to Delaval on the basis of such slim cost savings arguably was unwise. It was particularly unwise, the staff said, because on virtually all technical grounds the other two manufacturers were more qualified than Delaval to supply the generators. All three firms had experience manufacturing generators. However, Colt and Cooper-Bessemer had experience manufacturing generators for nuclear plants, and both had generators with a record of satisfactory use in nuclear facilities. Delaval's generators had no previous experience in nuclear applications, and the generators to be provided to Shoreham were considered prototypes (Department of Public Service Staff, 1984: 407-411). The DPS staff concluded that "LILCO purchased a prototype unit, with all its inherent risk, over two alternative proven units when there was no prospect of any benefit except a very small cost difference" (p. 411).

Second, there were very early indications that the diesels' crankshafts could not absorb the stresses of normal operations at Shoreham, but neither LILCO nor Stone and Webster responded effectively to the information. In 1974, Delaval made a series of calculations to determine the stresses put on the crankshaft and generator shaft by conducting a torsiograph test. Stone and Webster learned that the predicted stress levels exceeded the codes and standards set for Shoreham, and asked Delaval to actually test the stress levels put on the diesel engine. Delaval performed the test in 1975, but "the torsiograph test results provided by Delaval to Stone and Webster did not indicate the actual stress levels measured by the test...The torsiograph test report should have reported these stress values but did not...Stone and Webster should have realized the test report was incomplete and should have requested Delaval to provide this information. Without that information, Stone and Webster was not able to resolve its questions concerning whether the stress levels exerted upon the crankshaft would be in excess of allowable limits." LILCO's project engineer, in turn, "did not know what a torsiograph test was at the time...nor could he recall asking anyone to explain the torsiograph test...In fact, the company concedes that it did not know and never asked why Delaval was testing the crankshaft" (Department of Public Service Staff, 1984: 420-423).

There were other problems in the handling of the diesel generators. The relations between Stone and Webster and LILCO and Delaval were poor. Stone and Webster and LILCO questioned Delaval's quality control program, but did not intervene in it early enough or effectively enough. The diesels were not stored properly on the Shoreham site. After the diesels failed, LILCO's and Stone and Webster's handling of the generators, as well as Deleval's, were going to be scrutinized closely. This turned out to be one more instance in which an area of construction had not been managed effectively.

Disallowance of Costs

At the end of the prudence review the New York State Public Service Commission (PSC) disallowed $1.395 billion of Shoreham's then-estimated $4.62 billion cost. The Department of Public Service staff prepared a report which recommended an amount to be disallowed. The Public Service Commission itself ultimately decided on the disallowance after the DPS recommendations were reviewed in public hearings, and after its own additional analysis.

The disallowed amount was determined by comparing the performance of the Shoreham project to that of other nuclear projects. The Department of Public Service had earlier asked LILCO to analyze the effects of various factors on the costs of Shoreham, trying to specify how much of the cost could be linked to specific failures. However, the project's accounting and budgetary records were not clear enough to support any such analysis.

Because of the weakness of the financial records the DPS had to use indirect methods, and accordingly compared Shoreham's costs to those incurred at other nuclear projects, reasoning that excesses generally were due to mismanagement. There were four areas of disallowed costs--excess engineering hours, excess construction hours, costs of delays associated with mismanagement, and costs associated with repairing and replacing the emergency diesel generators.

To estimate excess engineering hours, the Department of Public Service took the average of three computations. First, it compared Shoreham's engineering hours to the engineering hours of the Susquehanna nuclear project, a project which also had a Mark II reactor system, received its construction permit seven months after Shoreham received its permit, was also located in the Northeastern United States (Pennsylvania), but was operating. This comparison implied 3.1 million excess hours. Second, it compared Shoreham's engineering hours to the average of a 21-plant sample of operating facilities; this comparison implied 2.7 million excess hours. Third, it compared Shoreham's ratio of engineering work time to construction work time to the ratios found elsewhere in the industry; this comparison implied 2.8 million excess hours. The DPS concluded that Shoreham incurred about 2.9 million excess engineering hours, costing $140.8 million.

To estimate excess construction hours, The DPS compared the direct construction work percentage at Shoreham (the share of construction time spent in direct work, which was about 21 percent) to the percentages which were found in three studies--a 26 plant study of the nuclear industry, another study of seven plants, and a study of a comparable nuclear plant. It then took the number of direct construction work hours actually incurred at Shoreham and subtracted from that figure the hours that would have been incurred if Shoreham's direct work percentage had been as high as the average work percentage reported in the three studies (Public Service Commission, 1985: 115-116). This generated an estimate of 26.3 million excess construction hours, valued--allowing for certain other costs--at $399.8 million.

The Public Service Commission, after examining the record, also estimated the costs of the construction delay which could have been avoided. It focussed on two separate delay periods--August 1982 through April 1984, and August 1984 to March 1986. The first period was related to construction delays, and the second to the need to replace the faulty diesel generators. The Commission estimated the costs of the first period by computing the costs of the time period associated with excess construction hours, and also compared Shoreham's construction schedule to the schedules of the comparable LaSalle and Susquehanna nuclear plants. Those plants were chosen because they had a similar design and, therefore, faced similar regulatory issues. The Public Service Commission regarded the 1984-1986 delay period as being due solely to the diesel generator problem.

The administrative law judges in the prudence hearing, and the Public Service Commission itself, concluded that actually only half of the delay could be reasonably blamed on LILCO, and that half reflected "the extraordinary [public] intervention and stringent regulatory supervision which adversely affected Shoreham's construction schedule" (Public Service Commission, 1985: 120). Accordingly, the PSC disallowed $759 million in costs associated with delays. Finally, the PSC disallowed $95 million in costs associated with repairing and replacing Shoreham's emergency diesel generators.

LILCO appealed the disallowances to the New York State Supreme Court, but in December the Court upheld the PSC's disallowances for excess engineering costs, construction labor costs, and diesel generator repair and replacement costs. However, the Court questioned the Commission's general conclusion that about half of the delay costs could be attributed to mismanagement and half to circumstances beyond LILCO's control:

> Unquestionably, LILCO's mismanagement and the antagonism to Shoreham's emergency planning are interrelated and concurrently generated these delays. But apart from the [Administrative Law Judges'] bald and conclusory declaration that "about half of these schedule delay costs were due to the special impact of Shoreham's unique licensing history and not to factors within the control of LILCO," there is no further elaboration in their decision or in the PSC's decision as to why these costs were, or should be, shared evenly...
>
> The relationship between LILCO's mismanagement and the resistance by Suffolk County and others, as causes of the delay, is obvious and demands further explanatory comment before it can be said that these delay costs have been fairly apportioned on a rational basis. At certain times one or the other may have alone been responsible for portions of the delay, and at other times the two causes coincided. Each seems to have had a ratchet effect on the other. Mismanagement delays gave opponents time to intervene and muster opposition, shook public confidence in the project and stimulated heightened administrative scrutiny, particularly that of the NRC and PSC. On the other hand, opposition-induced delays exacerbated the effects of LILCO's diffuse management practices, and Suffolk County's turnabout on emergency planning appears to have thwarted successful completion of a [required emergency response plan]. We realize that quantification of these catenating causes will not be easy, but something more than a visceral reaction to the evidence is

required (Long Island Lighting Company v. Public Service Commission, New York State Supreme Court, Third Judicial Department. Opinion 54452 (December 30, 1987), p. 14.

Accordingly, the Court--although upholding the disallowances for engineering and construction labor hours, and emergency diesel generator repair and replacement--told the PSC to specify the cases in which LILCO was primarily responsible for delay-related costs.

The Court reasoned that an exceptionally hostile political environment played a large part in delaying the plant in the 1980's. At that time the New York State and Suffolk County governments reversed earlier support--or at least tolerance--of Shoreham. That opposition was critical, because Federal law required a nuclear power plant to have a strong emergency preparedness plan in the event of an accident, and any such plan required state and/or local cooperation. The emergency planning regulations thus left open the possibility that state and local government could in effect veto the opening of a nuclear plant. The next chapter examines how the possibility materialized at Shoreham.

Appendix to Chapter 4
Chronology of Key Events

1965

April 13: LILCO announces plans to build the Shoreham nuclear plant, at a cost of $65-75 million.

1967

February 21: LILCO's Board of Directors authorizes the contract with General Electric to supply the turbine generator, nuclear steam supply system, and initial fuel load for Shoreham.

May 23: Edward Duffy, LILCO's executive vice-president, recommends the selection of Stone and Webster as architect/engineer and construction manager for the Shoreham project.

July 26: LILCO's Board of Directors approves the construction of Shoreham Unit 1, a 540 MW nuclear unit for commercial operation in the Spring of 1973 at a cost of $124 million.

1968

May 15: LILCO submits the Preliminary Safety Analysis Report (PSAR) for a 540 MW reactor at Shoreham to the Atomic Safety Commission.

October 1: LILCO vice-president Arthur Sugden recommends that the Long Island Lighting Company defer the Shoreham project in light of emerging concerns about nuclear plant construction.

November 21: LILCO notifies the Atomic Energy Commission that it plans to construct an 820 MW unit at Shoreham with a commercial operation date of May 1975. The Atomic Energy Commission agreed to retain the same docket for the amended PSAR for the 820 MW Shoreham unit.

1969

March 26: LILCO's Board of Directors authorizes the construction of an 820 MW unit at Shoreham at an estimated cost of $217 million, with a commercial operation date of May 1975.

April 21: LILCO files amended Preliminary Safety Analysis Report to reflect 820 MW Shoreham unit.

June: Arthur Wofford hired and named LILCO project manager for Shoreham.

July 1: Site clearing and grading begins at the Shoreham site.

September: LILCO increases estimated cost for Shoreham to $261 million, with projected commercial operation in May 1975.

November: Atomic Energy Commission proposes that all nuclear power plants receiving construction permits after April 1, 1970 be required to comply with ASME III standard, including B31.7 Piping Code.

December 15: The Advisory Committee on Reactor Safeguards issues a letter favoring the Shoreham construction permit.

1970

March 4: Stone and Webster recommends the use of the B31.7 code requirement for the design of piping systems at Shoreham.

August 24: Contract between Stone and Webster and LILCO for engineering and construction management services signed, effective retroactive to June 1, 1967.

September 1: Foundation work for turbine and reactor buildings begins.

September: LILCO increases the estimate of Shoreham's costs to $300 million with a projected commercial operation in May 1975.

September 21: Shoreham's construction permit hearings begin before a United States Atomic Energy Commission Atomic Safety Licensing Board.

1971

February 5: LILCO considers deferring the Shoreham project.

May: Atomic Energy Commission begins review of emergency core cooling systems in nuclear plants.

June 12: The Nuclear Regulatory Commission issues final regulations requiring plants with construction permits issued after January 1, 1971 to comply with ASME III, including B31.7 piping code.

July 23: United States Federal Court of Appeals issues *Calvert Cliffs* decision, requiring that the Atomic Energy Commission prepare environmental impact statements when considering licenses for nuclear plants.

August 3: LILCO instructs Stone and Webster to shut down engineering, design, and construction at the Shoreham project.

August: AEC regulatory staff issues an analysis saying that the proposed Shoreham emergency core cooling system is acceptably safe.

October 1: Stone and Webster halts Shoreham's construction.

November 4: Atomic Energy Commission Atomic Safety Licensing Board begins hearings on Shoreham's emergency core cooling system.

1972

January 31: LILCO increases its estimate of Shoreham's costs to $350 million, with an April 1977 commercial operation.

May 1: LILCO decides to reactivate engineering work for the Shoreham project.

June: Staff of Atomic Energy Commission issues Draft Environmental Statement for Shoreham concluding that construction permit is justified.

June: New York State Department of Environmental Conservation concludes that Shoreham can be built without violating State water quality standards.

September 27: LILCO's Board of Directors approves the selection of Stone and Webster for phase 1 of the proposed Jamesport nuclear project.

October 1: Construction and other work at the Shoreham site begins again after a one-year shutdown.

December 22: Atomic Energy Commission requests utilities to evaluate the effects in nuclear plants of high energy pipe breaks outside containment (PBOC), and provides criteria for preventing damage from such breaks.

1973

March 1: Thomas J. Burke becomes LILCO's project manager for Shoreham.

April 12: Atomic Energy Commission announces it is granting construction permit for Shoreham.

May 13: LILCO increases its estimate of Shoreham's cost to $560 million, with a commercial operation of July 1977.

June 14: First concrete is poured for the Shoreham reactor building.

October: First Arab oil embargo begins.

October 26: Atomic Safety and Licensing Appeal Board affirms the issuance of construction permit for Shoreham.

1974

February 28: Andrew W. Wofford reports that Stone and Webster cannot meet a target date of Spring 1978 for Shoreham's commercial operation.

March 26: LILCO's Nuclear Project Construction Group issues a "Self Analysis Report."

May 15: Charles R. Pierce becomes president of LILCO, and Andrew W. Wofford becomes LILCO vice-president of project management.

July 19: LILCO increases its estimate of Shoreham's costs from $506 million to $695 million, and changes target commercial operation from July 1977 to May 1978.

1975

January: United States Nuclear Regulatory Commission (NRC) assumes regulatory and licensing functions of Atomic Energy Commission (AEC).

January: To increase its direct contact with the project, LILCO moves the offices of its project manager Thomas J. Burke and the LILCO project team to the Shoreham site.

August 28: Final Safety Analysis Report and Environmental Report submitted to the NRC.

November 7: The Nuclear Regulatory Commission rejects as incomplete LILCO's final safety analysis report.

December 8: The reactor pressure vessel is set at Shoreham.

1976

January 15: Piping System Erection Progress Report issued, citing numerous problems at the project.

January 23: Final Safety Analysis Report and the Environmental Report accepted for review by the NRC.

May 1: LILCO and Stone and Webster form UNICO (Unified Construction Organization) as the Shoreham construction manager with

Stone and Webster in the lead role, and with direct participation by LILCO personnel.

June 30: LILCO increases its estimates of Shoreham's cost from $695 million to $969 million, and changes the target date of commercial operation from May 1978 to May 1979.

August 20: Piping erection task force is established with William J. Museler as chairman.

October 9: The first system at Shoreham is turned over to Start Up by construction.

December 7: Piping Erection Task Force Final Report issued.

December 20: Craft and Supervision Assistance Program (CASAP) final report issued.

1977

January 6: LILCO advises Stone and Webster that it would continue in lead construction manager role on a trial basis.

February 2: Nuclear Regulatory Commission rejects LILCO's approach to dealing with Pipe Break Outside Containment issues.

March 12: Stone and Webster advises LILCO that fuel load at Shoreham would be delayed between 15 and 28 months.

March 28: Arthur D. Little issues draft report to LILCO, discussing serious organizational problems at the Shoreham project.

July 5: Wilfred O. Uhl recommends that LILCO take over Shoreham construction management.

August 31: LILCO's Board of Directors approves LILCO's takeover of construction management at Shoreham.

September 12: LILCO assumes the lead role in UNICO and relieves Stone and Webster of construction management responsibilities.

September 14: LILCO agrees to comply with the intent of the Nuclear Regulatory Commission's rules on Pipe Break Outside Containment issues.

October: LILCO increases its estimate of Shoreham's cost to $1.223 billion, with a September 1980 commercial operation.

1978

January 1: Charles R. Pierce becomes Chairman of LILCO, and Wilfred O. Uhl becomes President of LILCO.

September 15: Joseph P. Novarro becomes LILCO's Shoreham project manager.

October 1978: LILCO's increases its estimate of Shoreham's cost to $1.337 billion, with a September 1980 commercial operation.

December 1978: Dr. William J. Catacosinos elected to membership on the LILCO Board of Directors.

1979

March 4: Three Mile Island accident occurs, resulting in a partial meltdown at the plant in Pennsylvania.

May 21: Public Service Commission begins investigation of the quality of management at the Shoreham project.

July 31: LILCO increases its estimate of Shoreham's cost from $1.240 to $1.581 billion, and changes expected commercial operation from September 1980 to May 1981.

1980

July: LILCO increases its estimate of Shoreham's cost to $2.213 billion, with an expected November 1982 commercial operation.

1981

July 1: William J. Muselar becomes LILCO's manager of construction and engineering at Shoreham.

July 24: LILCO increases its estimate of Shoreham's cost to $2.49 billion.

1982

June 5: Nuclear Regulatory Commission Atomic Safety Licensing Board begins Shoreham operating license hearings.

September: Pre-operational testing of the diesel generators at Shoreham begins.

September 30: Torrey Pines Technology issues a report on Shoreham, recommending changes costing approximately $1 million but finding plant's construction and quality assurance program to be satisfactory from the standpoint of safety.

1983

May: LILCO increases its estimate of Shoreham's costs to $3.3 billion, with expected commercial operation in April 1984.

June 8: LILCO files a motion with the NRC for a low power license for Shoreham.

August 12: Crankshaft on one of Shoreham's emergency diesel generators breaks.

October 24: Stone and Webster recommends that LILCO purchase 3 Colt diesel generators to replace the Delaval diesel generators.

November: LILCO orders diesel generators from Colt Industries.

1984

January 30: Charles R. Pierce resigns as LILCO chairman, and is replaced by Dr. William J. Catacosinos.

March 28: LILCO accepts the resignations of Wilfred O. Uhl, president, Thomas H. O'Brien and Charles J. Davis, senior vice-presidents, and Andrew W. Wofford, vice-president.

October: Atomic Safety Licensing Board issues low-power testing license for Shoreham.

December 21: The Department of Public Service staff files its initial brief on LILCO's management of Shoreham's construction. The report recommends that $1.9 billion of Shoreham's $4.2 billion cost to that point be disallowed because the cost resulted from poor management.

1985

August: Shoreham begins low-power testing.

December 16: The Public Service Commission announces that it will disallow $1.395 billion of Shoreham's costs as being imprudently incurred, about one-third of the plant's costs to that point.

1987

December 30: New York State Supreme Court upholds Public Service Commission's disallowance of excess engineering costs, construction labor costs, and diesel generator repair and replacement costs, but directs Public Service Commission to justify disallowance of delay costs in more detail.

Chapter 5

THE EMERGENCY PLANNING CONTROVERSY

The Strengthening of Planning Requirements

Since 1970 the Atomic Energy Commission (AEC), later the Nuclear Regulatory Commission, has recognized a need for procedures to protect the public in case of accidents at a nuclear plant. Prior to the Three Mile Island accident in March of 1979 emergency planning was not a major priority because the AEC/NRC considered the probability of a major accident to be so low and back-up safety systems to be so reliable. Applicants for an operating license had to indicate how they would respond to an emergency, but did not need to specify the steps to be taken to implement the plans. Also, the AEC/NRC evaluated emergency plans not when it reviewed the construction permit, but when it issued an operating license. This meant that an emergency planning problem might surface only after a multi-billion dollar facility was built (*Urban Law Review*, 1980).

The Three Mile Island accident changed this. The worst nuclear accident in the United States up to that time occurred on March 4, 1979 at the Number 2 nuclear unit at General Public Utility's facility in Harrisburg, Pennsylvania. The accident was actually a series of events. Pumps supplying feedwater to steam generators closed down. Emergency pumps automatically began working, but certain valves had been closed mistakenly, and these prevented water from reaching the generators. The containment system should have then sent additional water to the generator, but another valve stuck open and the cooling water was sent to the floor of the containment building rather than to the generator. A generator boiled dry, and then operators mistakenly shut down additional pumping systems designed to supply emergency cooling water. Consequently the reactor boiled dry, failed to cool, and began to disintegrate. The sequence stopped before the core fully melted, but demonstrated the possibility of a severe nuclear accident (Tomaine, 1987: 1).

Although the damage was contained on-site, there was a hastily-organized emergency response in surrounding communities. According to former NRC Commissioner James Asselstine,

> There had been little planning by state and local governments responsible for dealing with the emergency, and the response was confused. There were no procedures for coordination among various governments; there were no clear lines of authority; there were no clear procedures for or means to disseminate information; there were no clear procedures for determining whether to take protective action or how to carry it out once it had been decided upon; and few if any of the other elements essential to an effective emergency response existed. Because of the disarray on the part of nearly everyone involved in the response to the TMI accident, people living in the area around the plant did not know what information was accurate and did not know whether it was safe to stay in the area or whether to leave. Most people simply did whatever they thought best (52 Federal Register 6985 [March 6, 1987]).

The ad hoc nature of the response indicated that emergency planning had to be formalized and strengthened. A General Accounting Office Report, written even prior to the accident, criticized the state of preparation for accidents at nuclear plants (U.S. General Accounting Office, 1979). A report to the Nuclear Regulatory Commission, and the Presidential Commission on the accident--the Kemeny Commission--also concluded that preparations for nuclear emergencies had to be improved greatly (Nuclear Regulatory Commission, 1980; Presidential Commission on Three Mile Island, 1980).

In December 1979, the Nuclear Regulatory Commission proposed new emergency preparedness rules, and adopted final rules in the Summer of 1980 (10 C.F.R. 50.47, and 10 C.F.R. Part 50, App. E.). Applicants for operating licenses had to have plans that would provide "reasonable assurance that adequate protective measures can and will be taken in the event of a radiological emergency." Emergency plans had to specify how utilities and county and state governments would coordinate their activities in case of an accident, and how they would alert and advise the public about the accident (The NRC considered county government to be the primary local government unit, and towns, cities, and municipalities to be sub-organizations of counties; see U.S. Senate Committee on Environment and Public Works, 1983: 51). An applicant had to demonstrate that facilities and personnel to carry out the plan would be available. Plans were to cover an area of about 10 miles around the plant, and 50 miles in radius covering the path of emissions from the plant. Evacuation of ten miles around the plant were to be planned in the event of a worst-case accident, although evacuation was only one possible type of emergency response. The Federal Emergency Management Agency (FEMA) and the Nuclear Regulatory

Commission both played roles in evaluating emergency plans. FEMA reported its findings to the NRC, and the FEMA findings created a "rebuttable presumption" for the NRC that the plan was or was not adequate (45 *Federal Register* 82713 [December 16, 1980]; U.S. Senate Committee on Environment and Public Works, 1983: 62).

State and local governments could design a plan that was more protective than Federal regulations required--for example, they could establish a wider emergency planning zone. However, the NRC would follow its own standards when making regulatory decisions, and not the standards of the state or local governments. For example, the Nuclear Regulatory Commission could approve a nuclear plant with a ten-mile Emergency Planning Zone (EPZ), even though the local government felt that a wider zone was desirable (U.S. Senate Committee on Environment and Public Works, 1983: 50-51, 60).

Although the Federal government regulated the safety of nuclear plants, and had the authority to decide whether the emergency plan for a nuclear plant was adequate, it was expected that state and local governments would take the lead in actually *implementing* emergency plans. As Shattuck writes,

> Predicating the acceptability of such necessary emergency plans on state and local participation was logical and practical. The interest of local governments in the protection of the exposed populations was likely to be strong and their familiarity with local conditions great. No purely federal program could effectively take into account the myriad local differences which shape the performance of information, transportation, distribution, and law enforcement tasks during emergency management. No matter how large or well-trained a federal emergency task force, no matter how well-equipped to direct emergency operations or how clearly authorized to help cajole or coerce frightened or angry people, it was unlikely that federal assistance could be rendered in time. In the worst case, where mass evacuation became necessary within a matter of hours, if not minutes, the central role of local authorities and the support of state personnel would be indispensable. Without the assistance of local officials, without their detailed knowledge of schools, nursing homes and hospitals, of traffic patterns and water supply lines, no plan--no matter how credible on paper, how detailed and comprehensive--could be effectively implemented. An emergency plan, said one NRC commissioner, would "not be worth anything unless local authorities commit [sic] to supporting it" (Shattuck, 1986: 236).

The law therefore presented a striking issue. While the Federal government exclusively evaluated emergency plans for licensing purposes, adequate plans presumably required state and local cooperation. This created a situation in which a state or local government conceivably could "veto" a proposed Federal operating license for a nuclear plant simply by not cooperating in the emergency preparedness planning. The utilities were expected to provide a plan which involved state and local participation, but the utilities obviously had no authority to *require* state and local participation.

The NRC said in 1980 that it might grant a nuclear plant a license even if an emergency plan had some deficiencies (45 *Federal Register* 55,406 [August 19, 1980]). For example, a utility might be able to compensate for a weakness in state or local performance: "[I]n determining the sufficiency of adequate interim compensatory action" the NRC would examine "State plans, local plans, and licensee plans to determine whether features of one plan can compensate for deficiencies in another plan" (45 *Federal Register* 55,403). There was little need for compensatory action up to the early 1980s, however. Emergency planning operated relatively smoothly in most states, with state and local governments developing emergency plans jointly with utilities. The locality would generally take the lead, consistent with their traditionally central role in emergency response, and then forward the final plan to the state government for approval. After its approval the state would forward the plan to the Nuclear Regulatory Commission. There the NRC would ask the Federal Emergency Management Agency to review the plan, and run a series of on-site tests. FEMA gave its evaluations to the NRC as part of the operating license review.

Occasionally local governments had refused to work with utilities in planning, and in those cases state governments had worked directly with the utilities. One such case occurred in New York in 1983. Rockland County objected to the draft emergency response plan for the Indian Point nuclear facilities in Buchanan, New York. Although the County agreed to participate in any *actual* emergency response, it would not participate in any *exercise* of the plan required by FEMA and the NRC. Instead, the County--one of four falling in the emergency planning area--proposed to develop a plan with which it would be satisfied. This left the Indian Point plant with an emergency plan without the cooperation of one of the area's counties, a stalemate threatening to shut down the plant. New York State then moved in to replace Rockland County in this stage of the planning process, and the NRC subsequently voted 3-2 to allow Indian Point to continue operating (Rockland County rejoined the planning effort in January, 1984). In other cases, such as the Diablo Canyon plant in California, the State had not approved a plan but the County had worked directly with FEMA (*New York Times*, February 19, 1983: 26; May 15, 1983: XXI-1; June 10, 1983: 1; January 18, 1984: II-4).

The Shoreham situation was unique. In 1983 the government of Suffolk County formally refused to participate in emergency planning, and Governor Mario Cuomo neither tried to change the County's decision nor had the State prepare the plan directly. As discussed earlier, the 1979 emergency planning regulations created the chance that state and local governments might try to veto a nuclear plant's operating license by blocking emergency planning. The veto materialized at Shoreham.

The County's Split with LILCO

Early Cooperation

As discussed in Chapter 2, Shoreham's construction permit hearings from 1970-1973 had been contentious. The Lloyd Harbor Study Group (LHSG) objected to the plant for many reasons, the difficulties of emergency planning being only one. In its petition to intervene in the hearings, LHSG said in March 1970 that "If an accident occurred at... Shoreham, appropriate and effective arrangements could not be made...to protect the public health and safety and to permit ready removal and evacuation of people...from the project area and nearby population centers" (Cited in U.S. Senate Committee on Environment and Public Works, 1983: 33). LILCO argued that the LHSG objection was out of order, saying that the AEC had to address emergency planning at the operating license stage and not at the construction permit stage. The AEC's staff, the Atomic Safety Licensing Board (ASLB), and the ASLB Appeals Board all agreed with LILCO. The Appeals Board said that "The final determination with respect to the content of the necessary emergency plans must depend on the development of the population growth in the next few years...The Board correctly concluded that the outline of the proposed emergency plan submitted by the applicant is sufficient at this time" (6 *A.E.C.* 831, 851 [October 26, 1973]).

Unlike the Lloyd Harbor Study Group, the Suffolk County government did not oppose Shoreham at the construction permit hearings. At the first session of the hearings County Executive Lee Dennison told the Atomic Safety Licensing Board that "I urge you, as County Executive, to grant immediate licensing for construction of this nuclear power facility as proposed at Shoreham." His final report to the County Legislature in 1971 urged "[u]nited, aggressive, stand-up-to-be-counted action by the governments of Suffolk County" supporting Shoreham. His successor, John Klein, noted to the Legislature in January 1972 that "My predecessor has openly and aggressively endorsed the proposal for the Shoreham Atomic Plant. In this annual report, I reconfirm that executive commitment and support" (cited in U.S. Senate Committee on Environment and Public Works, 1983: 128).

Some County officials expressed concern about emergency planning when the *operating* license proceedings began in 1977. A County petition to intervene in those proceedings claimed that LILCO and the NRC staff "have not prepared and assessed an adequate evacuation plan for Shoreham as required...with regard to the timing and feasibility of evacuation including, but not limited to, consideration of road networks, water barriers, weather conditions, and population density" (August 1, 1977; cited in U.S. Senate Committee on Environment and Public Works, 1983: 337).

Still, up to 1981 Suffolk County was developing an emergency plan cooperatively with LILCO. In February 1973 County Executive John Klein ordered his Department of Emergency Preparedness to develop a response plan for "radiation incidents." In 1975 and 1976 Suffolk County officials met with LILCO regarding emergency planning roles and responsibilities. In 1977 the County Executive ordered the County Departments of Emergency Preparedness and Transportation to develop a plan for Shoreham; County Executive Klein, and then the New York State Office of Disaster Preparedness, approved that plan in 1978.

After the Three Mile Island accident the Federal government strengthened the emergency planning regulations. In December 1979, Klein and LILCO signed a memorandum of understanding updating the County's responsibilities in emergency planning. LILCO later suggested that the County use a private consultant--to be paid by LILCO--to develop a new plan. In February 1981, the County decided to develop the plan itself, rather than use a private consultant. LILCO still agreed to support that arrangement, and in March signed a contract with the County under which LILCO would pay the County $245,000 to develop a Radiological Emergency Response Plan within six months.

The Suffolk County Legislature, which had not been consulted on the contract, refused to allow the County to accept the initial contract payment. LILCO revised the contract, and the Legislature approved it in September 1981. LILCO issued a first check to the County for $160,000, with the balance of the payment due when Suffolk would presumably submit a final plan in March 1982.

But in February 1982 Suffolk County essentially split with LILCO over emergency planning. Suffolk had retained a Washington law firm to represent it in the upcoming Shoreham operating license hearings, and one of its attorneys, Herb Brown, recommended that the County cancel the emergency planning contract with LILCO and return the $160,000 paid to date because the County would be an intervenor in the operating license hearings. On February 19, 1982, the Suffolk Planning Director told LILCO that the County was returning the money paid to date, and was withdrawing from the LILCO contract.

In March, LILCO told Suffolk that it was "relying on the County to perform its obligations under the contract." The County Legislature, however, passed a resolution which authorized the County to hire its own consultants, out of its own funds, to prepare another emergency plan by October 1982. The Suffolk County consultants would include professors of planning, sociology, geography, and psychiatry; the behavioral scientists were added to consider the possible response of residents and emergency workers in the event of an emergency. Suffolk argued that its new planning effort would also evaluate the effects of Long Island's geography in any emergency in more detail than the original Planning Department document, and would include an assessment of the probabilities of various types of accidents (*New York Times*, May 12, 1982: II-2; May 16, 1982: XXI-15). That plan would have to be approved by the Legislature before being submitted to New York State, the NRC, and FEMA (Suffolk County Resolution 1266-82, March 9, 1982).

In April, LILCO asked the New York State Disaster Preparedness Commission (DPC) to review the original County-LILCO plan for forwarding to the Federal government, and the State DPC agreed. LILCO submitted the plan one month later, noting in a cover letter that although Suffolk County agencies had prepared the plan, the County government itself had not endorsed it. Ira Freilicher, a LILCO Vice-President, said that the plan was more than adequate, that LILCO wanted to start training its employees for their roles in emergencies, and that further delaying the opening of the plant could cost millions of dollars. LILCO's President, Charles Pierce called the plan--that he very much wanted approved--a "superior piece of work."

The County government immediately threatened legal action to stop State approval of the plan. It called the County Planning Department document "resource material" only, and said that the only legitimate plan being developed was its own, expected to be ready by October. The next week the County legislature established a subcommittee to investigate whether LILCO violated any laws by submitting the document to the State. (In November of that year a County attorney concluded that LILCO had not because of the cover letter's note that the County had not endorsed the plan.)

On June 2, 1982, the State Disaster Preparedness Commission noted a number of problems with LILCO's plan, and said that it would be willing to review any revision. However, the DPC Chairman also noted that the draft was the "best we have received from any county in the state, and there is no fundamental reason why the plan cannot be corrected fairly easily." On June 17, LILCO Vice President Ira Freilicher met with Deputy County Executive Frank Jones to see if Suffolk County was willing to resume work with LILCO on emergency planning. The County was not. In September another LILCO vice-president wrote to the County, saying that "Let me close by repeating that much of the work LILCO has done on the offsite plan, as well as the resources and experience of LILCO employees and consultants, would be of use to the County in developing its plan. LILCO takes no pleasure in working on emergency planning without the County. I urge you to resume joint planning with LILCO as soon as possible" (Letter from Matthew Cordaro to Chief Deputy County Executive John Gallagher, September 17, 1982). Over the next months LILCO continued to try to get Suffolk to work with it on emergency planning; the County opted to continue to work separately.

The County's Withdrawal from Planning

In July 1982, Suffolk County announced the results of its consultants' survey of public responses to a nuclear emergency. The study concluded that 25 percent of Long Island residents would try to leave the Island in the event of an accident at Shoreham, even though they had been told not to do so. The Suffolk Deputy County Executive asserted that one of the implications of the survey was that the County would have to plan to manage the behavior of people in a larger area than specified by the Nuclear Regulatory Commission planning requirements (*New York Times*, July 16, 1982: II-2).

Suffolk County released its consultants' draft emergency plan early in December (Herr and Associates, 1982). The plan specified evacuation of a five to seven mile radius around the plant in the event of a major accident. The plan also specified that six other "outer sectors"--covering up to a radius of twenty miles--would have to be cleared depending on several factors including wind direction at the time of the accident. The plan projected accident consequences as follows:

To estimate the number of persons potentially injured or killed in the event of an accident at Shoreham, the consequence of the more severe of two likeliest (1 in 10,000 years) core melt accident categories was estimated by simulating 100 such accidents under different contingencies of release characteristics and weather. In most of the simulated cases, there were no early fatalities at all, in a few cases there were 400-600 early fatalities, in one case 1,600 early fatalities. Averaged over all of the simulated accidents, 25 early fatalities resulted if no protective action were taken, and 15 early fatalities if the entire affected population were to shelter themselves. Similarly, there were early injuries affecting up to 2,600 persons without sheltering, 1,300 with it, with an average of 67 persons injured if no protective action is taken, 18 persons if the population takes shelter.

Longer term health consequences result from latent cancer induced by radiation exposure, even at very low irradiation dose levels. For the same category of accident and assuming "worst case" wind towards New York City, accident-induced latent cancer fatalities within the 20-mile zone range from none at all, given sufficiently rapid evacuation (more rapid than feasible here) to 1,800 fatalities, based on a 3-hour departure delay and 20 travel hours to clear the zone, a performance which this plan should be able to improve upon. Beyond the EPZ the dose levels and the resulting incidence rates are much lower, but the potentially affected population is much higher. The result is 7,000 latent cancer deaths outside the EPZ if such an accident occurs, added to whatever number results from protective actions or their lack within the EPZ (Herr and Associates, 1982: 357).

The plan projected that evacuating the inner area of 5-7 miles around the plant would take 3 3/4 to 5 1/4 hours, depending on various factors; the entire EPZ would be evacuated in between ten to twenty hours (Table 12).

The plan noted that the projected evacuation times "may prove to be optimistic...The traffic model doesn't reduce road capacity to allow for breakdowns (running out of gas in a 10-hour traffic jam will be commonplace), for aberrant behavior (drivers may not behave in an orderly fashion after several hours in what they believe to be a life-threatening queue, in need of gasoline, water, food, or toilets), or for demand in excess of free-flow capacity (some similar models reflect reduced capacity or through-put when demand greatly exceeds capacity)." "On the other hand," the document continued, "the model also assumes no increased flows attributable to people using break-down lanes, shoulders, and opposite-direction lanes. It also assumes that everyone evacuates westward, which for some persons is irrational, and which therefore is probably over-conservative" (Herr and Associates, pp. 349, 351).

As the County was announcing its draft plan in December, the State Disaster Prepardedness Commission (DPC) was about to approve the LILCO-submitted plan for forwarding to the Nuclear Regulatory Commission. Suffolk County asked the New York State Supreme Court to grant a temporary restraining order preventing the DPC approval until the County could complete the final version of its plan. On December 7 the Court issued the restraining order. Within a week, at a Court hearing, all parties agreed that Suffolk would have to submit a final plan by February 22, 1983, or forfeit the right to oppose LILCO's plan.

In January the Suffolk County legislature held hearings on its plan. During the hearings a majority of the Legislature appeared opposed to the plan and to Shoreham itself, although the Presiding Officer of the Legislature, Louis Howard, supported both. The opposition became official in mid-February. On February 16, after the hearings concluded, Suffolk County Executive Peter Cohalan said that he did not believe that it was possible to prepare an adequate evacuation plan at all. He noted that Shoreham was sited in 1970, when an offsite response plan was not a key element of the planning process, and argued that the narrow geometry of Long Island and its enormous traffic problems made evacuation virtually impossible. (A State Fact Finding Panel on Shoreham, which divided on the merits of opening Shoreham itself, nevertheless concluded that "the Shoreham plant is located on a site which the Shoreham panel believes today would be found unacceptable for a nuclear power plant" [(New York

Table 12. Evacuation Times Projected by Consultants To Suffolk County (in hours)

Sectors Recommended for Evacuation	Time for 1985 Summer Population	Time for 1985 Winter Population	
		Normal Cond.	Adverse Cond.
Inner Only	4:30	3:45	5:15
Inner + Western	14:30	13:45	22:15
Inner, Western, and WSW	19:15	17:30	28:30
Inner, Western, WSW and SW	21:15	19:45	30:00
Inner, WSW, SW and S	21:15	19:45	30:00
Inner, South, SE and East	17:30	10:45	17:15
Inner, SE, and East	15:00	10:45	17:15
Inner and East	12:30	4:30	6:45

Source: Selected from Herr and Associates, 1982: 350. Table in original contains additional scenarios.

State Fact Finding Panel, 1983: 8)]. Cohalan said that "the unique local conditions of Long Island make it impossible to protect the public safety if there was a serious accident at the Shoreham plant. There can never be emergency preparedness to protect our citizens. The result is that Shoreham should never operate" (*New York Times*, February 18, 1983: 1). Later he added that "For me there is only one choice... Suffolk County should reject the draft county plan and any other plan by bluntly stating this inescapable conclusion: it is impossible to establish emergency preparedness in Suffolk County that can protect the public health, safety, and welfare in the event of a serious nuclear accident at Shoreham."

One day after Cohalan's announcement Governor Mario Cuomo ordered the State not to approve any evacuation plans for Shoreham that had not been cleared by the Suffolk County government. That week the Governor himself had expressed reservations about Shoreham. After a questioner alluded to the risks of nuclear power indicated by the Three Mile Island accident and the periodic problems at the Indian Point facility in Buchanan, New York, he said "Shoreham is another good example. I've taken a position on Shoreham regularly...over the last few years. Nuclear power is a threat. I think I have said that if there is not clear proof that Shoreham or any of the other plants are safe, and if there is not clear proof that you have an evacuation plan that is safe and workable, then there ought to be no nuclear facility" (*New York Times*, February 18, 1983: 1). Cuomo said that he would "not be a party to any effort to impose an independently developed state plan on Suffolk County," although he would "cooperate in any way possible" with LILCO and Suffolk County to work out arrangements for an effective emergency plan, if possible (*New York Times*, March 3, 1983: II-1).

Hours after the Governor's order, the Suffolk County Legislature voted 15-1 to cease emergency planning efforts on the grounds that it was impossible to protect all Suffolk County residents in the event of a nuclear accident, and that presenting any plan "would be misleading to the public by indicating to County residents that their health, welfare, and safety are being protected when, in fact, such is not the case" (Suffolk County Resolution 1196-83, February 23, 1983). The next week Cohalan and the Suffolk County Legislature asked the Nuclear Regulatory Commission to halt its Shoreham operating license hearings because there would be no County support for the mandatory emergency planning, and "[w]ithout such preparedness the N.R.C. could not find the requisite reasonable assurance that Shoreham can be operated without undue risk to public health and safety" (*New York Times*, February 19: 26; February 20: XXI: 1; February 24: II-2).

Some actors sharply criticized the County's action and Cuomo's support of it. LILCO had invested over $3 billion in the plant to date, and Chairman Charles Pierce claimed that abandoning the plant would threaten to turn Long Island into an "economic wasteland" (*New York Times*, March 11: II-1). Nassau County Executive Francis Purcell said that abandoning the plant could have "potentially disastrous economic consequences" for Long Island; Nassau County later threatened to sue Suffolk County if Suffolk's refusal to cooperate in emergency planning kept the plant idle, increasing electricity rates on the Island (*New York Times*, February 26: 26; March 13: XXI-1). A *New York Times* editorial said that while it was known retrospectively that building the plant was a mistake, the present costs of abandoning Shoreham could force LILCO into bankruptcy and increase Long Island's electric rates--the third highest in the nation--even further, and that Shoreham's risks did not warrant its abandonment (February 27: IV-16).

In May 1983 the Nuclear Regulatory Commission voted 4-0, with one Commissioner abstaining, to consider LILCO's emergency plan even in the absence of Suffolk County and State cooperation. The NRC said that "as we read the applicable regulations, we are not only authorized but also obligated to at least consider any proferred utility offsite emergency plan. Further, while there may well be serious issues of federal preemption involved in the current offsite emergency planning controversy, we find it unnecessary to reach such issues at this time because, as we read the applicable regulatory provisions, the agency is obligated to consider a utility plan submitted in the absence of state and local government-approved plans and has the ultimate authority to determine whether such a submission is sufficient to meet the prerequisites for the issuance of an operating license" (NRC Memorandum and Order CLI-83-13, May 12, 1983). Commissioner Victor Gilinsky abstained because the Commission arguably had avoided the question of whether any emergency plan lacking state and local cooperation could ever be judged adequate. He said that the answer to that question was "clearly, no...however they may qualify their views now, I do not believe that a single Commissioner would actually approve the operation of the plant without such participation" (cited in U.S. Senate Committee on Environment and Public Works, 1983: 82). In October, the Federal Emergency Management Agency accepted the referral of the plan from the Nuclear Regulatory Commission for review (*New York Times*, October 6, 1983: II-2).

The Marburger Panel

By 1983 Shoreham was almost complete. However, Suffolk County's refusal to participate in emergency planning, and Cuomo's decision to back the County, conceivably blocked the opening of the plant. Thus, in 1983, the issue of whether an almost $4 billion dollar facility would ever open at all was posed sharply for the first time.

On April 19, 1983, shortly after backing the County, Governor Cuomo proposed the creation of a "Fact Finding Panel" on Shoreham. The Panel was to examine Shoreham's economics, safety, and policy issues. Perhaps most important, it was to advise the Governor on whether or not the State ought to intervene to replace Suffolk County in emergency planning. The Panel was chaired by Dr. John Marburger, President of the State University of New York at Stony Brook (located about twenty miles west of Shoreham).

There was little agreement in the Panel's December 1983 Final Report regarding Shoreham's economics, safety, and emergency planning. According to Marburger's transmittal letter, the Final Report contained a core of findings "which most Panel members support to some extent," at least according to members' individual interpretations of wording. The members disagreed on major issues, however. Thirty-seven pages of the Final Report were devoted to the issues around which there was something of a consensus, and 144 pages to the divergent supplementary views of individual members or coalitions of members. In effect, the Panel's experience was a microcosm of the politics of Shoreham in the early 1980s.

One early conception of the Fact Finding Panel was that it was would a small, technically specialized group. The actual Panel was broadened to include representatives from both Suffolk and Nassau counties. Its final members included some individuals not allied with either side of the Shoreham issue, and others strongly committed to either completing or abandoning the plant. The members were:

> Dr. John Marburger, President, State University of New York at Stony Brook (Chairman);
>
> Dr. David Axelrod, Commissioner, New York State Department of Health;
>
> Karen Burstein, at the time of appointment the Executive Director of the State Consumer Protection Board;
>
> Leon Campo, the Assistant Superintendent of the East Meadow Schools and, as Chairman of the People's Action Coalition, a leader of the groups opposing Shoreham (appointed in June, 1983);
>
> William Dircks, Executive Director of Nuclear Operations for the Nuclear Regulatory Commission;
>
> Marge Harrison, co-Chairperson, Long Island Progressive Coalition and Vice-Chairperson, New York State Democratic Committee;
>
> Dr. Alfred Kahn, of AEK Associates and Cornell University, and former Chair of the New York State Public Service Commission;

Dr. Herbert Kouts, Chairman, Department of Nuclear Energy, Brookhaven National Laboratory (endorsed by the Citizens for Orderly Energy Policy--primarily a group of Brookhaven scientists favoring the plant--and appointed in June);

Dr. Paul Alan Marks, President, Memorial Sloan-Kettering Cancer Center;

David McLaughlin, Deputy Associate Director of State and Local Programs and Support, FEMA;

William Ronan, C.C.X. Inc. and former chairman of the Metropolitan Transit Authority (Nassau County representative);

David Wilmott, Editor and Publisher, Suffolk Life (Suffolk County representative);

Hugh Wilson, Institute for Suburban Studies, Adelphi University.

The political legitimacy of the Panel became an issue after the initial appointments, and Cuomo responded by adding members. The Governor appointed Leon Campo in June to counter the charges of anti-Shoreham groups that the Panel was tilted in favor of opening the plant. In turn, Herbert Kouts was added to counter the charges of others that Campo would tilt the Panel in favor of abandoning the plant. Also, Marge Harrison, a strong opponent of the plant, expected to be traveling during some of the meetings and requested permission from the Governor to designate an alternate. David Sprintzen of C.W. Post College was the alternate selected. Campo, Harrison, Willmott, and Sprintzen opposed the opening of Shoreham strongly. Kouts was from Brookhaven National Laboratory, and had been endorsed by scientists who favored opening the plant. Other members had no conspicuous leaning either way as the Panel began its work. Staff from the New York State Consumer Protection Board, State Energy Office, Office of the Governor, SUNY-Stony Brook, State Department of Health, State Public Service Commission, and the Nuclear Regulatory Commission and the Federal Emergency Management Agency, were also selected to prepare technical materials for the Panel.

The Panel discussed, in addition to safety and economics, whether or not the State ought to intervene to assure that an emergency plan was prepared for Shoreham, in effect overruling Suffolk County.

The Issue of Plant Safety

Radiation Risk

The Nuclear Regulatory Commission and Environmental Protection Agency had estimated the levels of radiation exposure which ought to trigger emergency responses. These levels were called "Protective Action Guides," or PAGs. The PAG for radiation exposure for the general population was 5 rems (roentgen unit man) of radiation (United States Nuclear Regulatory Commission and Environmental Protection Agency, 1978).

The Shoreham Panel staff presented an estimate of the excess mortality that might result from an accident exposing the public to ten rems of radiation. The source of the estimate was the National Academy of Sciences (NAS) Report of the Committee on Biological Effects of Ionizing Radiation (called the BEIR III report). BEIR III figures suggested that one million people exposed to ten rems of radiation would experience between 766 and 2,255 more cancer deaths than the 163,800 cancer deaths that would normally occur in the population.

After the BEIR III report had been published, Dr. Edward Radford, Chairman of the BEIR III Committee (and later a consultant to Suffolk County), reported that more recent measurements of data from the Hiroshima bombing required an increase in these mortality estimates. He said that recent measurements indicated that a figure of 4,000 additional deaths per million population, rather than 766-2,255 deaths per million, was more accurate. The Shoreham Panel Staff noted that "Although the BEIR III estimates remain the 'official' risk estimates endorsed by the NAS, it does appear likely that future estimates may double based on the new dosimetry used in evaluating the Japanese data."

The Staff noted that other estimates of health effects were much higher than the BEIR III figures or even the higher figures suggested by Radford, writing that a small number of scientists had called for "increasing 'official' estimates by factors of twenty or more" (Panel members opposing Shoreham frequently referred to these much higher figures). The Panel Staff commented that "There does not appear to be significant support for [these higher estimates] among health physicists or radiation biologists" (New York State Fact Finding Panel [subsequently cited as Fact Finding Panel] Staff Report, p. 3[a]8). Thus, the generally accepted data suggested that a major Shoreham accident exposing the public to ten rems of radiation would induce up to 4,000 deaths per million population. Other figures were much higher, but were not supported by most scientists.

The Likelihood of Major Radiation Release

Both LILCO and Suffolk County consultants estimated the likelihood of a major accidental release of radiation at Shoreham. These estimates attempted to factor in the characteristics of the plant, the prevailing weather conditions of the area, the history of accidents at other nuclear plants, and other considerations. Although both sets of consultants incorporated such considerations, any final estimates were necessarily highly uncertain (The Panel Staff Report discusses the LILCO and Suffolk County analyses in detail on pp. 3[a] 32-56).

LILCO's consultants estimated that five rems of radiation would be exceeded at ten miles once in 400,000 years. Suffolk County's consultants estimated that five rems would be exceeded at ten miles once in 6,000 years. Differences in technical assumptions produced the sixty-six fold difference in results. The differences included differing treatments of how the reactor was or was not designed to absorb radiation, or of the likelihood of operator error. Also, the Suffolk County consultant's risk estimate included an estimate of radiation release that was recognized to be an overestimate; the consultant indicated this reservation in his report to the County. The Staff Report to the Fact Finding Panel noted that the use of the higher figure "was not a trivial point" (p. 3[a] 42).

The NRC, commenting in a preliminary way on both risk assessments, said that "Results quoted by both LILCO and Suffolk County from the Shoreham probabilistic safety study do appear to be plausible. Note, however, that the state-of-the-art in predicting reactor risk is not such that precise projections can be made, either of accident likelihood or of consequence. It is common to find that reactor risk assessments somewhat underestimate accident likelihood and overestimate accident consequences" (cited in Staff Report, p. 3[a] 56).

The consensus section of the Panel's Final Report did not take an official, strong position on the relative safety of Shoreham. This is not surprising because the issue sharply divided panelists. The Report did discuss three relevant issues--whether there was a scientific consensus on the impact of low-level radition, whether LILCO was able to manage the plant safely, and whether the plant itself was physically safe.

Panel's Conclusions Regarding Low-Level Radiation

On October 1, 1983, the Panel met in an executive session to consider the conclusions of the Report. At that meeting, some members stressed that the Report had to recognize the existence of a majority scientific view on the degree of risk from low-level radiation.

[SPEAKER NOT IDENTIFIED]: Jack [Commission Head John Marburger], I would suggest that we are never going to agree on the significance of the incremental risk associated with radiation exposure from the operation of a nuclear plant...There are those of us who feel that normal operation in the absence of incidents does not contribute significantly to the overall risks of the general population. On the other hand, there are those who take a strong position that any additional radiation is going to have a long-term adverse impact, and we really do not have a basis for completely rejecting either of the hypotheses...

KOUTS:...there is a weight function that should be attached to this. On the one hand, we have...all these large bodies of scientific opinion who investigate and come to conclusions on the meaning of low-level radiation. We have monitoring activities at all the nuclear power plants to determine what comes up, and the total of this leads to a certain set of conclusions at one end of the spectrum. At the other end of the spectrum I can [only] name four people.

AXELROD:...I would suggest...that we say something along the following lines: The clearest exposition of the controversy over the impact of low-level radiation occurs in the context of BEIR III in which the minority of credible scientists have taken the position that it is "X," as opposed to the majority position which is "Y," with respect to the long term impact of low-level radition. BEIR III is a credible document.

One of the panelists, not identified in the executive session transcript, maintained that the debate over low-level radiation in fact was not settled:

It is extremely important that if you are going to characterize the views, and if you want to add Herb's [Kouts] note about where the weight [of scientific information] is at present, you have to add two things. You have to add, number one, the historical continual change in the lowering of what is considered a legitimate dose, and you have to also add the fact that there is a significant controversy now which leads in the direction of suggesting that the radiation risk is significantly underestimated (Transcript of October 1, 1983 Panel Meeting [subsequently referred to as Transcript], pp. 2-3).

Although the discussion focussed on the risks of normal operations, the positions paralleled those taken on the relative health risks of accidents. Referring to the risk of low-level radiation, the consensus section of the Final Report noted that "Because of the low levels involved, and the difficulty of distinguishing among the various possible causes of actual cancer, the numbers are somewhat controversial. The best available study estimates the increases in lifetime risk to be from .5% to 1.4% of the naturally occurring mortality rate for each ten rems of exposure" (Fact Finding Panel Final Report, 1983: 7). The study referred to was BEIR III. In introducing the consensus section of the Report, Marburger did allude to the existence of majority and minority views on the question of risk: "Conclusions supported by the overwhelming majority of scientists are given little public credence if it appears that the scientific enterprise itself relies for its support upon agencies whose aims may be advanced by a particular finding. In this atmosphere of suspicion, minority scientific views may be accorded much more credence by the lay public than the majority views of the scientific community" (Final Report, p. 3).

Panel's Conclusions Regarding LILCO's Ability to Manage the Plant Safely

The Final Report's consensus section expressed strong reservations about LILCO's ability to operate Shoreham safely:

> LILCO had not adequately prepared itself for its venture into nuclear power. It may well be that no utility has adequately prepared itself for such a high technology operation. It is in its attitude toward safety in general, and the associated meticulous approach to safety related operations that the nuclear power industry bore much criticism by the Three Mile Island Commission. One has the impression that the utilities, at least the one we have been studying, have struggled to keep up with the initiative seized by the NRC. They seem to have been in a reactive mode wherein adherence to NRC guidelines replaced independent initiative....Several Panel members have had experience in the administration of large and complex organizations. We recognize in some of the complaints about inefficiency and incompetency the inability of individuals in the chain of command to appreciate the larger sweep of their work and the inevitable effects of complexity on daily operations. Whether these explanations suffice for the vast uneasiness expressed by so many of our witnesses must be answered by a different mechanism than our Panel. We urge the Public Service Commission to move its investigation of imprudency in Shoreham construction costs as rapidly as possible (Final Report, p. 22).

Panelists generally shared this view, but drew different conclusions. Leon Campo's supplementary statement said that "LILCO continues to act in a manner which radiates its insensitivity and misunderstanding of the potential for danger inherent in the operation of a nuclear power plant" (Final Report, pp. 49-50). In contrast, Alfred Kahn, Herbert Kouts, and William Ronan agreed that "LILCO did not prepare itself adequately to be an operator of a nuclear plant, and that it still lacks credibility as an operator of a nuclear plant...We believe, however, that the deficiency can be remedied by one device or another--whether by LILCO itself or, at the extreme, by bringing in a more experienced operator--before start up of the plant" (Final Report, p. 62).

Panel's Conclusions Regarding Construction Quality

Aside from the issue of whether LILCO was competent to manage the plant's operations, the Fact Finding Panel examined the issue of whether the design and construction of the plant were of sufficiently high quality to merit confidence in its operation.

Inspecting the plant was one way to assess Shoreham's design and construction quality, and by 1983 the Shoreham plant had been inspected several times by several organizations. However, Suffolk County wanted an inspection over which it had direct control. LILCO and Suffolk County could agree neither on the scope of such an inspection nor on the degree of follow-up work that inspectors could require. Suffolk County maintained that LILCO wanted to avoid possibly extensive but necessary repairs and design changes. LILCO, in turn, maintained that Suffolk County and one of the County's consultants, MHB Associates, were inclined to require excessive changes and thus delay the plant further, substantially increasing its cost and possibly killing the project. LILCO was particularly concerned because MHB Associates was the consultant; MHB's partners had resigned from General Electric, designer of the Shoreham reactor, and had testified before Congress that the reactor's design was flawed (Transcript, p. 9).

In 1982, LILCO hired Torrey Pines Technology to review the previous inspections of the Shoreham plant. Torrey Pines recommended changes which cost approximately $1 million, but in general found the plant's construction and its quality assurance program to be satisfactory from the standpoint of safety. After the Torrey Pines review, Suffolk County legislator Gregory Blass requested that three independent experts in nuclear physics and engineering further evaluate the plant's design and construction quality (Herbert Kouts was one of the individuals selected; the Shoreham Panel had not yet formed). LILCO indicated that it was willing to have Suffolk County's consultants--including MHB Associates--join the inspection, but Suffolk refused to sanction the arrangement because it did not feel that its consultants would have adequate authority to do follow-up inspections or to require additional work on Shoreham. Nevertheless, the three-man panel reviewed the Torrey Pines report and concluded that "Torrey Pines Technology's verification was a careful and honest job, and we agree with their overall conclusions." The review panel also examined the NRC inspection report, particularly with respect to complaints about the quality of construction, and concluded that "they seem to have done a competent and believable job of investigating the allegations made to them" (Final Report, p. 78; Transcript, p. 10).

At the Panel's October 1 meeting, John Marburger, in trying to identify points on which panelists could agree, said that "...there are allegations of construction problems...I would be willing to say that there is widespread public perception of construction problems...there is very intense public feeling that maybe the construction is inferior" (Transcript, p. 8). A speaker, who was not identified and was possibly a member of the staff rather than a Panel member, commented that

SPEAKER NOT IDENTIFIED: I'm a little concerned. Clearly you can write whatever you want, but I think you have an obligation when the facts are there to have an element of balance. Here's an example: question--I agree with David's characterization of what the people on Long Island believe to be true in terms of the construction. On the other hand, there have been, as I understand it, three separate reviews of the design of some of the systems in construction, and each one of them to varying degrees--the Torrey Pines, the Teledyne one, and the NRC itself, concluded essentially that the design and construction that those three groups looked at was adequate. I'm not qualified to say whether they were right or wrong. I don't think, with the exception of probably two people here, anybody else is, but findings are factual information, and I think you need them in the report. It's that type of balance I'm trying not to lose.

Marge Harrison responded that the inspection process of the NRC had proven itself to be suspect:

HARRISON: I mean, at what point does society begin to learn something. We see what happened at Diablo Canyon, where virtually they were ready to pull the switch, and it had been inspected, etc. etc., and...somebody pulled a document off a shelf and found there was a major problem. At what point do we indicate to the Governor, that the very regulatory monitoring processes have proven themselves [to be inadequate]. The NRC's own admission with regard to what they did with the revocation of Diablo Canyon's license, has shown that there is a major problem that the agency has to address. That's a very important thing. To take a superficial view of the regulatory process is absurd (Transcript, p. 8).

Another panelist questioned the integrity of the inspections that had been completed at Shoreham:

SPEAKER NOT IDENTIFIED: It is important to talk about those inspections and why, on the one hand, there are reasons for people to be concerned about those specific ones...Suffolk County...insisted upon a full inspection design review and its participation [but such an inspection] never came about. That is a fact. It never came about. LILCO did conduct those inspections...I think we have to reflect what's on the record as to why people question the integrity of what happened in those inspections (Transcript, p. 8).

Another panelist noted that:

SPEAKER NOT IDENTIFIED: These are the facts. Not only did LILCO work with MHB, they agreed at one point, with MHB and the County, to four system inspections. Where it broke down was when the full definition of inspection and design, which remains to this day a problem with that company as to why it won't open its doors, was finally defined as 32 operating systems, and you'll find in the Torrey Pines report as to what we really wanted, what the County felt was needed, and what LILCO pretty much agreed was the definition. Where the problem came in is that from the fore, what they hoped would take place is that the County would select two out of four systems. If there was nothing found in those two, there was no reason to go beyond that. That was the kind of approach taken. As opposed to that, the insistence was that all 32 operating systems be looked at, and if MHB made a determination for the County, that one system needed to have additional work or inspection done, that they would not be vetoed, either by LILCO...well, by LILCO.

HARRISON: There wasn't any agreement on what the threshold would be. In other words, the County's problem was at what point, if they had a suspicion, would they be permitted to follow that up...(Transcript, p. 10).

Marburger questioned the inference that LILCO was trying to avoid inspections of the plant out of concern for its quality of construction:

MARBURGER: Now I have to say something about that. I asked a lot of questions in that particular meeting about exactly what happened and what people's attitudes were. My conclusion is that the County's technical objections to the Torrey Pines inspections are not critical. That is to say, the County's technical objections are of very little weight. They're legitimate disagreements about the technical field of how you sample. I think on the whole, I would have to come down more on the side of the plant, and the design, than on the County consultant's approach, which was too blind because it didn't take advantage of special knowledge. Regarding the split between the County and the lack of faith between the two when it came right down to the participation of the County, I don't understand that as anything particularly nefarious. I see it as a direct result of the fact that LILCO was afraid that the consultants of the County were going to deliberately delay the project, and they were just as conscious of the "time is money" equation then as they are now. The consultants that the County wanted to have were people who had resigned from G.E., and who testified before Congress... regarding the failures of G.E. design. These were the people that the County wanted to essentially have veto power over the details of the inspection. LILCO took one look at that and they said, "These guys are going to hold us up" (Transcript, p. 9).

The Final Report did not express a consensus view on the adequacy of the plant's design and construction. It did note that

A subsequent limited review of the Torrey Pines study by a team assembled by a County Legislator indicated that the methodology used by Torrey Pines was sound. The Torrey Pines study focused on a limited number of systems of the Shoreham plant and made recommendations to LILCO for changes which cost approximately $1 million. The NRC has also conducted a number of its own inspections of portions of the Shoreham plant. The difficulty with the inspections to date are primarily their limited nature and the exclusion of participants who are seen to represent the public interest other than the NRC, e.g., Suffolk County (Final Report, p. 10).

In their supplementary statements, some panelists maintained that, by any reasonable standard, the plant could be considered safe. Other panelists maintained that the costs of a worst case accident, and the questionable trustworthiness of LILCO, the nuclear power industry, and government regulators, implied that the plant probably was not safe and should not operate. The split between the two groups of panelists reflected fundamentally different views about the legitimacy of the available information and of industry and government regulation in general. This point will be discussed further below.

The Issue of Shoreham's Economic Value

The estimated costs of operating versus not operating Shoreham varied greatly. Whether Shoreham was an economic asset or a liability depended on factors like the future price of alternative energy sources, future demand for energy, and other variables.

The consensus section of the Final Report noted that analyses by its own staff, LILCO, and Suffolk County consultants all concluded that "There is an economic advantage, although possibly small, to operating the Shoreham plant versus not operating it." However, the estimated penalty of abandoning Shoreham varied substantially. LILCO estimated that abandonment would cause electric utility rates to rise an additional eighteen percent, Suffolk estimated a 2.2 percent increase, and the Panel staff estimated an eight percent increase (Final Report, pp. 29-30). These analyses did not factor in additional delays or problems with the plant, such as the problems with the emergency diesel generators which developed in August 1983. The analyses also were sensitive to long range assumptions about oil prices and other variables. Thus, the panelists' supplementary statements differed greatly on any incremental economic benefits of operating Shoreham. The anti-Shoreham panelists maintained that there actually would be an economic benefit to abandoning the plant (Final Report, pp. 54-57, 62-64).

All panelists agreed that Shoreham would have a major economic impact because the plant had been built and would have to be paid for in one way or another. The Final Report's consensus section noted that "projections of the incremental cost of Shoreham operation versus abandonment are far less than the projected economic consequences of Shoreham regardless of whether it operates or not. This is to say that the average annual electric rate differential of approximately 8 percent per year is dwarfed by the over 50 percent short-term increases projected under conventional rate treatment regardless of whether the plant operates" (p. 34). However, the statement by Alfred Kahn et. al. said that even though the incremental cost of abandonment was "far smaller than the probably $4 billion capital cost, it is nevertheless a very big number" (Final Report, p. 63).

The Issue of State Intervention in Emergency Planning

The most immediate policy issue facing the Panel was whether any economic advantage of operating Shoreham warranted action by the State to ensure operation of the plant--specifically, whether New York State should prepare directly an emergency response plan if Suffolk County refused to do so.

The Report commented that Suffolk County had commissioned studies "of reasonable quality." The Report also noted that the County's decision to refuse to participate in planning "is a result of governmental, not purely scientific or technical, processes. That is, the County asserted its right to make decisions of government based upon whatever information its legislative and executive branches have available to them. The distinction between the County consultants' studies and the actual position taken by Suffolk County was stressed by the County and seems significant to the Panel as well" (p. 35).

Suffolk County's consultants also had distinguished between the report to Suffolk County and Suffolk County's policy-making. In hearings on the draft emergency plan, consultant Philip Herr noted

> I am satisfied that the response plan is just about as good as could be devised for this circumstance, and that the consequence estimates are fully supportable by scientific analysis.
>
> Even though the plan is about as good as can be achieved, the outcomes may not be satisfactory to you. The risk is not and could not be reduced to zero. In the event of a severe plant accident, there are likely to be substantial public health consequences, not because of any emergency plan failure, but because that plant at that location makes those consequences essentially unavoidable.
>
> Deciding whether or not those consequences are acceptably remote and small in relation to other considerations is not our charge. Rather, we want only to make clear what the consequences resulting from an accident might be given implementation of the best plan we can devise (cited in Senate Committee on Environment and Public Works, 1983: 296).

As discussed earlier, the consultants' report estimated the consequences of "the more severe of the two likeliest (1 in 10,000 years) core melt accident categories" by simulating 100 such accidents under different contingencies of release characteristics and weather.

In most of the simulated cases, there were no early fatalities at all, in a few cases there were 400-600 early fatalities, in one case 1,600 early fatalities. Averaged over all of the simulated accidents, 25 early fatalities resulted if no protective action was taken, and 15 early fatalities if the entire affected population were to shelter themselves. Similarly, there were early injuries affecting up to 2,600 persons without sheltering, [up to] 1,300 injuries with it, with an average [over the simulated accidents] of 67 persons injured if no protective action is taken, 18 persons if the population takes shelter.

In addition, "assuming a 'worst case' wind towards New York City, accident-induced latent cancer fatalities within the 20-mile zone range from none at all, given sufficiently rapid evacuation (more rapid than feasible here), to 1,800 fatalities, based on a 3-hour departure delay and 20 travel hours to clear the zone, a performance which this plan should be able to improve upon." The study had estimated a time of 3 3/4 - 5 1/4 hours to clear the five to seven mile inner zone, and 10-30 hours to clear the outer rings, depending on the scenario. Assuming a worst case accident and the worst case wind toward New York City, the consultants estimated 7,000 deaths *outside* the emergency planning zone because of the New York City population (Herr and Associates, p. 357).

According to some of the panelists, these conclusions, even if accepted, suggested that the odds of such a major accident were extremely remote; that the odds of the worst case scenarios in particular were infinitesimal; and that the risks assumed were similar to those assumed in the use of other technologies such as other types of energy. These panelists also pointed out that the consultants had not concluded that emergency planning was not feasible, but *only that Shoreham's risks could not be entirely eliminated*. The supplementary statement by Kahn, Kouts, and Ronan observed that

The County's consultant did not conclude that an emergency plan was impractical or impossible. The County, acting contrary to the studies which it commissioned, concluded that there might be some emergency situation of extremely small probability under which they could not guarantee the safety of everyone in Suffolk County. Therefore they would not participate in any emergency plan.

This action involves a remarkable exercise of governmental discretion--a refusal to participate in planning for an emergency, presumably in the expectation that by so refusing, they could remove the threat itself (Fact Finding Panel Final Report, p. 61).

However, granting that one might disagree with the technical bases of the County's decision, the Panel's conclusion essentially deferred to the County's authority to make such a decision, provided that the decision was not clearly "frivolous" or harmful to other geographic areas:

If this Commission were to conclude that the action of Suffolk County, in refusing to prepare an off-site emergency preparedness plan, was frivolous or carried with it severe economic consequences for the people of Suffolk County or other jurisdictions outside of Suffolk County, then a recommendation for State intervention would clearly be appropriate. As discussed previously, the Commission does not view Suffolk County's action to be frivolous. Furthermore, the Commission has been impressed that the projections of the incremental cost of Shoreham operation versus abandonment are far less than the projected economic consequences of Shoreham regardless of whether it operates or not. This is to say that the average electric rate differential of approximately 8 percent per year is dwarfed by the over 50 percent short-term projected increases projected under conventional rate treatment regardless of whether the plant operates. Based on information which has been presented by the Company, the County, and the staff, this Commission has concluded that such a recommendation for State intervention to prepare an emergency plan is not warranted (Final Report, p. 34).

It is important to realize that the statement does not suggest that the Suffolk County decision was (or was not) the correct decision. Rather, the Panel reasoned that there was a presumption that Suffolk County had the right to make decisions on the facility if the County bore most of the consequences of the decisions. According to the Report, unless the County made a decision that was clearly unreasonable and thus a disservice to its residents, or that affected other jurisdictions substantially, the decision should be allowed to stand. The Panel's consensus neither endorsed nor repudiated the County's decision, but rather concluded that the decision was not so unreasonable as to warrant intervention by a higher level of government.

One objection to the conclusion, not recorded in the consensus section of the Report, was that the consequences of the decision indeed did spill over the borders of Suffolk County. One panelist noted that "...we have to recognize that a decision in this matter extends beyond Suffolk County. In other words, this impinges upon the whole nuclear regulatory activity and gets into the question of whether you can have a local veto of a federal policy" (Transcript of October 1 Panel Meeting, p. 7). Similarly, Kahn, Kouts, and Ronan commented that

The Report says that, assuming Suffolk County's decision on emergency planning was not reached in a capricious or frivolous manner, and assuming that the financial effect of not operating the plant as opposed to operating it would not bankrupt LILCO, neither of these two factors would be a basis for state intervention in the licensing process. In fact, there are other possible reasons to argue for state intervention in easing licensing and operation of the plant.

One is the strength of the nation, as it requires energy independence. A decision not to complete Shoreham--in effect, to discard a $3.5 or $4 billion essentially completed plant, on the basis--in our view unjustified--adduced by Suffolk County, would go far to foreclose the nuclear power alternative for both New York State and the country. We regard such an action as highly imprudent, for both the State and the country (Final Report, p. 43).

Impact of the Fact-Finding Panel

The Panel was established to present Governor Cuomo with facts on Shoreham's economics, safety, and policy issues. However, it could not arrive at a consensus on the safety of the plant or the costs of operation versus abandonment because its members viewed the scientific and technical information about nuclear power so differently. (Thus, Marburger's transmittal letter to the Governor referred to the Panel's efforts "to achieve, if not consensus, at least the mutual recognition that further efforts to achieve consensus would bring diminishing returns.")

Differences in political views shaped the results more than the ambiguity of the information itself. For example, estimates of low-level radiation hazards did vary, but not so greatly as to warrant substantially different policy conclusions. The BEIR III Report indicated that ten rems of radiation would lead to about 2,255 additional cancer deaths per million (a 1.4 percent increase over the expected cancer mortality) while Edward Radford's revision estimated that actually about 4,000 additional deaths would occur (a 2.4 percent increase). This was a difference, but not a qualitatively major one. Similarly, the policy implications of the different risk assessments arguably did not differ greatly. The LILCO consultant estimated that a major accidental release of radiation would occur once in 400,000 years. In one of the Suffolk County reports, a consultant estimated a risk of a major accident that was 66 times greater--once in 6,000 years--although the latter estimate was known to be high. Yet, 1 in 6,000--the admittedly high risk estimate--was still a relatively small probability of a major accident.

Those favoring the opening of Shoreham perhaps might have been willing to accept even the higher figures on radiation risk and accident likelihood. On the other hand, those opposing Shoreham were not doing so on the grounds that a 2.4 percent increase in cancer mortality in the event of a major accident, or a one in 6,000 chance of such an accident, were too great. Rather, their major reason for opposing Shoreham was the theoretical possibility that an accident might lead to Long Island incurring

40,000 early deaths, 35,000 delayed deaths, and a quarantine for twenty years (Willmott and Harrison Supplementary Views in Final Report, Section V-D, pp. 26-28).

Thus, there were two different approaches to scientific and technical information in the Shoreham Panel debate. Some panelists gave the information great weight. Because the plant was so controversial, rational individuals taking this approach could still insist that the information lean heavily toward one conclusion before they would favor opening the plant. Yet, if one accepted the scientific process, the evidence that Shoreham was reasonably safe was substantial. The available information did not suggest that the plant presented clearly unacceptable risks. The plant had been extremely costly, as discussed in Chapter 4, but each of the inspections of the plant's design and construction had been generally favorable. Testing had uncovered problems, such as the diesel generator or pressure suppression issue in the Mark II design, but that was the purpose of the testing.

Panelists opposing Shoreham rejected the information provided by these inspections and tests as irrelevant because they rejected the legitimacy of the organizations conducting them. Starting from a critical background, panelists could cite specific events confirming their suspicions--the Three Mile Island (TMI) accident, the problems with the Diablo Canyon construction, the problems with Shoreham's diesel generators, and so forth. Panelists *favoring* the opening of Shoreham maintained that any large scale construction project will have design flaws, elements of mismanagement, and other problems, and that Shoreham's critics were mistakenly inferring from not terribly surprising problems that the nuclear industry was distinctively untrustworthy, hazardous, and mismanaged (Transcript, p. 7; Final Report, p. 77). Panelists *opposing* Shoreham saw such events as instances in which industrial and government efforts to conceal a general pattern of such problems from the public momentarily unraveled. Thus, opponent Leon Campo noted that

> The record of the Shoreham Commission underscores the validity of the criticism that the NRC is primarily in the business of promotion of nuclear power and secondarily with public protection. For the Federal Emergency Management Agency to set aside its normal review of evacuation plans in favor of a Shoreham "Quickie"...is exemplary of the Federal Government's attitude toward safety...The current process of government, particularly those concerned with the New York State Public Service Commission and the Federal Nuclear Regulatory Commission do not warrant the continuation of the public trust...The citizens must conclude that the NRC views its responsibility to be solely one of licensing nuclear power plants which includes modifying the rules as necessary to accomplish same. The function of government in magnifying the public good is undermined by these self-serving, transparent initiatives which serve only the vested interests...(Final Report, pp. 40-41).

Regarding LILCO, as noted above, Campo said that "LILCO continues to act in a manner which radiates its insensitivity and misunderstanding of the potential for danger inherent in the operation of a nuclear power plant" (p. 49).

Operating from this critical perspective, Shoreham's opponents also cited the testimony of scientists who had broken ranks with, or had resisted, what they called an orchestrated scientific consensus. In a key testimony, Dr. Michio Kaku, Professor of Physics at the Graduate Center of the City University of New York, testified regarding what he maintained were unique dangers of Shoreham. Kaku's testimony allowed Shoreham's critics to maintain that "It has been statistically projected by physicists that...in a core meltdown coupled with steam explosion...the containment building ceiling could be blown apart and result in an uncontrolled spewing forth of radiation" (Willmott and Harrison, op. cit., pp. 25-26). Suffolk County legislator Wayne Prospect and panelists also relied on Kaku's testimony in a press release charging that "The Shoreham reactor is more dangerous than Three Mile Island" (Press Release from Wayne Prospect, Suffolk County 15th District, October 3, 1983).

The controversy over Kaku's testimony indicated just how far apart Shoreham's opponents and proponents were. Although anti-Shoreham panelists aggressively publicized Dr. Kaku's claims, other Panel members and some Panel staff called his claims misleading. On September 21, 1983, Herbert Kouts wrote to Chairman Marburger that "Following our meeting of the Shoreham Commission on September 14, it was suggested that I should join Dr. Kaku in further discussion of reactor safety matters before some or all of the Commission. I am writing to inform you that I will not do this. I find Dr. Kaku's comprehension of nuclear power safety to be so flawed that it would not be worth my time. He revealed too many misconceptions as to the scientific questions involved" (Letter from Kouts to Marburger, September 21, 1983).

Disagreements on technical issues were only part of this exchange. A greater part was the question of the credibility of organizations and individuals. To Kouts and others favoring Shoreham, scientific and technical information from a variety of sources generally pointed to a conclusion that the plant should open. To Kaku and others opposing Shoreham, this scientific "consensus" had been orchestrated by those with political and personal stakes in nuclear power.

Recognizing these differences, the Final Report's consensus section did report one major conclusion. The Panel concluded that the Governor should not intervene in emergency planning because to do so would deprive Suffolk County of its governmental prerogative to make non-frivolous decisions that primarily affected citizens within its own borders. The consequences of that conclusion were important. Ever since Suffolk County terminated emergency planning Cuomo had shown no strong inclination to intervene, and the Fact Finding Panel supported that decision. After that point LILCO probably would not receive any help from New York State in opening Shoreham.

LILCO'S Tax Maneuver and Cohalan's Switch

On January 10, 1984, one month after the Marburger Report had been released, LILCO withheld a $26 million Shoreham property tax payment due to Suffolk County, and put the money into an escrow account. LILCO argued that in the past Suffolk County had assessed Shoreham as if it were an operating plant, and thus as a more valuable facility than one under construction and not yet in service. LILCO also maintained that "Not only has Shoreham been assessed and taxed as an operating plant, but the level of Shoreham's assessments has run over 53 percent above the average assessments on other parcels in town." Since 1977 LILCO had paid $139 million in property taxes on the plant, and it was seeking a $58 million refund. Putting the $26 million tax payment into the escrow account would assure that the government would have a large part of the $58 million refund available should the company win the appeal (*New York Times*, January 11, 1984: II-4).

Furthermore, LILCO claimed that it would be entitled to an even larger refund if Suffolk succeeded in blocking Shoreham because the plant's value would be reduced to salvage value. A LILCO spokeswoman said that the company had researched the legal issues for over a year, and that the move "will help insure that area municipalities have the money to refund the company if it wins a pending appeal of its tax assessment." Another representative commented that "We would welcome the opportunity to meet [Suffolk County] in court over this issue. What the county appears to be crying crocodile tears over is exactly what would happen if it prevailed in its attempt to keep the plant from opening...It's just inconsistent for the county on one hand to say 'Abandon it,' and have the other hand outstretched for millions of dollars." LILCO added that it would pay the tax when the County decided to join the emergency planning efforts (*New York Times*, January 11, 1984: II-4; January 12, 1984: II-2; January 15: XXI-6; September 27, 1984: II-6).

The County government argued that the move was clearly intended to "put a gun to the head of the Shoreham-Wading River School District" and the County over Shoreham. The superintendent to the school system called the effects of any such loss in taxes "a catastrophe. I suspect we'd be bankrupt. I don't know how soon." Frank Jones, the Deputy County Executive, rejected the idea that LILCO had any right to withhold taxes, as the money had essentially been provided by State regulators to finance its operations. He also argued that the company had proceeded with the project at its own risk, and thus the County had no obligation to pay back any taxes if the plant failed to open.

In May 1984, the County Legislature's Budget Review Office released a study which said that property taxes in Suffolk County would have to increase seven percent--and sixteen percent in the area around Shoreham--if the plant failed to open and be taxed as an operating unit. The study commented that "A healthy, vital Suffolk may fall victim to its own policy of stopping Shoreham" (*New York Times*, May 27, 1984: 52). In June, County officials agreed that the Shoreham tax problem--added to the County's sewer district tax problem--could leave Suffolk with a shortfall of $200 million in a total budget of $800 million, including Federal and state aid. County comptroller Joseph Caputo, who had opposed County Executive Peter Cohalan on other issues, said that Cohalen and his aides had "painted themselves into a corner" on Shoreham, and that the policy would "bankrupt this County" (*New York Times*, June 24, 1984: XXI-1). In November bankers lending money to the county raised concerns about the taxes withheld by LILCO, and one County legislator opposing Shoreham also began questioning the continued legal and consulting costs of Suffolk's efforts to block the plant (December 2, 1984: XXI-4).

LILCO thus turned Suffolk County's effort to block Shoreham into a potential fiscal liability for the County. At the same time, some emergency planning regulations themselves began to be questioned. Estimates of the radiation that would be released following a major core melt accident were based on a 1975 reactor safety study, authored by Norman Rasmussen, which relied on small-scale experiments and computer simulations. The NRC regulation that required emergency planning to cover an area of about ten miles around the plant were based on those earlier estimates; at the time Suffolk County was arguing that evacuations of twenty miles around the plant might be necessary.

The 1979 Three Mile Island accident allowed scientists to estimate radiation release using an actual accident. By 1981 a number of individuals, and the Nuclear Regulatory Commission in June, began suggesting that the earlier study had overestimated the size of the probable radiation release in a core melt accident. The Three Mile Island accident indicated that a large amount of radiation that was expected to leave the reactor was contained in solid form within the reactor. In November 1984 Federal Energy Secretary Donald Hodel argued to Peter Cohalan that the radiological danger from a nuclear accident was "much less than most people once thought," and that a two mile evacuation zone around the plant might be "more appropriate" than a ten mile zone. Lee Koppelman, the Suffolk County Planning Director who developed its original ten mile evacuation plan, said that a five mile evacuation was "eminently doable" (*New York Times*, December 2, 1984: XXI-1).

Ever since February, 1982, Cohalan had led the fight against Shoreham on the grounds that the magnitude of an accident, and Long Island's special conditions, made it impossible to evacuate the Shoreham area. The emerging argument that it might be possible to reduce emergency planning zones, coupled with the County's financial problems from LILCO's withholding of taxes, started to erode Cohalan's opposition to emergency planning. On May 31, 1985, Cohalan reversed his position, ordering County employees to test an emergency response plan for the plant. In a move directly linked to Cohalan's announcement, LILCO announced that it had paid $131 million in back taxes to the County. Cohalan argued that he changed his mind based on the new evidence that a prudent evacuation around Shoreham was possible. Suffolk's fiscal problems, and LILCO's offer to stop withholding disputed taxes in exchange for County cooperation, undoubtedly contributed heavily to Cohalan's new support of emergency planning (*New York Times*, June 9, 1985: XXI-5).

However, the eighteen person Suffolk County legislature--sixteen members of which opposed the plant--complained strongly that Cohalan had no authority to reverse unilaterally the County's opposition to Shoreham. In May 1982, the County Legislature had passed a resolution, signed by Cohalan, which said that "Suffolk county shall not assign funds or personnel to test or implement any radiological emergency response plan for the Shoreham Nuclear Plant unless the plan has been approved, after public hearings, by the Suffolk County Legislature and the County Executive" (Suffolk County Resolution 1516-82, May 19, 1982). Citing that resolution, members of the legislature said that they would contest Cohalan's order in court. Governor Cuomo also continued to oppose the plant (*New York Times*, June 1, 1985: 1, 26).

On June 10, a state judge nullified Cohalan's executive order as requested by the County legislature, a decision later upheld on appeal (*New York Times*, June 11, 1985: II-2; June 25, II-4; July 10: 1). After losing an appeal of the decision, Cohalan issued a modified Executive Order which indicated that he would ask County employees "from time to time...to investigate, gather information, review, and evaluate radiological emergency response plans," instructions which stopped well short of actually cooperating in any implementation or test of a plan (Suffolk County Executive Order Number 2, July 1985). After County legislators opposing Shoreham were re-elected in November 1985, Cohalan said that "As long as the Federal government adheres to the rule of law and its own rules, I think the Shoreham nuclear power plant is dead." Although he said that he personally favored a drill to determine whether a safe evacuation could be done, Cohalen would oppose any Federal test of an emergency plan that did not include local participation in the test. Such a test occurred in February, 1986.

The Federal Test Of Emergency Planning

Suffolk County first withdrew from emergency planning in May of 1983. In April 1984, the Federal Emergency Management Agency told the NRC that it was planning, in a preliminary way, an exercise of the emergency plan drafted by LILCO, to be carried out by Federal and LILCO employees. A FEMA official said that "We have a mission to do emergency planning around nuclear power plants that can't change because an official, even a governor, says something." A Suffolk official called the idea of such an exercise "preposterous." The secretary to Governor Cuomo said that "It sounds like a 'let's pretend' kind of action by the Federal Government and a private company to exercise police power, which we don't think is legal or desirable in terms of assuring safety" (*New York Times*, April 28, 1984: 26).

On June 20, 1985, the Nuclear Regulatory Commission formally asked the Federal Emergency Management Agency to conduct "as full an exercise...as is feasible to test offsite preparedness capabilities at the Shoreham Nuclear Power Plant." Samuel Speck, the Associate Director for State and Local Programs for FEMA, responded that while such an exercise could be conducted, "the reluctance of county and State officials to participate in such an exercise and the related legal authority issues would place special parameters on the conduct of a LERO (Local Emergency Response Organization) plan."

> Any exercise without participation by State and local governments would not allow us sufficient demonstration to reach a finding of reasonable assurance [that the plan would protect the public safety]. This conclusion is based on the current legal decisions with respect to utility authority to perform civil emergency functions. However, that does not preclude the conduct of an exercise that would provide an indication to the Nuclear Regulatory Commission (NRC) as to utility onsite and offsite emergency capabilities. We believe such a report would have value in decisions to continue the licensing process or possibly provide a basis on which the NRC could make predictive findings. Obviously, the value of such an exercise in the licensing process is a determination which can only be made by the NRC (letter from Samuel Speck to William Dircks, NRC, October 29, 1985).

Some key members of Congress, including the chairs of the Senate and House Committees in charge of FEMA's appropriations, had also contacted FEMA regarding the emergency planning issue. A letter from Senator Jake Garn and Edward Boland to FEMA said that the budgetary Conference Committee was

...concerned that the emergency planning process for nuclear facilities may unnecessarily delay the operation of new facilities, threaten continued operation of currently licensed plants and involve unnecessary expenditure of Federal funds and personnel time by FEMA. In particular, the Committee is concerned about situations in which State or local government entities arbitrarily refuse to develop radiological emergency preparedness plans or to participate in the exercise or implementation of such plans. The Committee does not believe that the State and local government entities should be permitted to veto the operation of commercial nuclear facilities simply by refusing to participate in the preparation, exercise, or implementation of such plans.

The letter said that FEMA should presume that state and local governments would respond in the event of an emergency; that FEMA should provide the resources necessary to have adequate emergency plans in place; and that FEMA had the authority to step in and take over emergency planning where state or local governments refused to participate (letter from Senators Garn and Boland to Director Julius Becton of FEMA, November 13, 1985. Senate Alfonse D'Amato of New York had been the only dissenter in the Conference Committee vote).

On November 14, 1985, FEMA confirmed that it would conduct a test of the emergency plan, with or without State and local cooperation. Governor Cuomo called the decision "a complete reversal of Federal policy and a violation of the sovereignty of New York State and Suffolk County" (*New York Times*, November 14, 1985: 33). In January, 1986, the Suffolk Legislature approved a bill that prohibited any emergency preparedness drill for Shoreham without prior review and approval from the Legislature. The law threatened anyone who participated in the drill--namely Federal employees and consultants, and LILCO personnel--with criminal charges (January 12, 1986: XXI-1; January 14: II-5; January 19: XXI-1; February 8: 30). LILCO sued in Federal District Court to overturn the law. Peter Cohalan called the LILCO suit "the last dying quack of a lame duck, a ludicrous and irresponsible lawsuit designed to intimidate county officials in their legitimate exercise of their duties to provide for the public safety." Wayne Prospect, a Suffolk County legislator, called it "a sham lawsuit" (January 26, 1986: XXI-I). However, on February 10 a Federal judge barred Suffolk County from trying to block the exercise of LILCO's emergency plan, arguing that the County law "impermissibly intruded" on Federal regulatory authority (*Long Island Lighting Company v. County of Suffolk*, 628 F. Supp. 654, 666).

On February 13, LILCO, the NRC, and FEMA ran the test of the emergency response plan. The test was started when FEMA described an emergency situation to a LILCO official at the plant, and continued through the afternoon. By early afternoon the test had simulated the evacuation of 138,000 people from the area around the plant. Over one hundred Federal observers were on-site to evaluate the plan and LILCO's performance (*New York Times*, February 14, 1986: II-2).

An aide to Governor Cuomo reiterated that "This exercise was truly meaningless" because Suffolk County and the State had not participated. Also, the Regional Director for FEMA, Frank Petrone, said at a subsequent press briefing that FEMA could not give a "reasonable assurance" that the plan would protect the public safety because the exercise had not assessed how well the affected governments could implement the plan, although he added that "We feel that here at Shoreham the state and local government will respond if there is an emergency" (*New York Times*, February 16, 1986: 48).

However, on April 14, Petrone resigned from FEMA over the wording of the official Post Exercise Assessment. Petrone said that for several years he had felt that Suffolk and New York State's refusal to participate in Shoreham's planning undermined any possible plan, and that FEMA had felt "pressure" to weaken emergency planning requirements to expedite Shoreham's licensing. In the final Assessment, he wanted to include the statement that "Since the LERO [Local Emergency Response Organization] plan cannot be implemented without State and local participation, FEMA cannot give reasonable assurance under NUREG-0654 that the public health and safety can be protected." When he insisted that the statement be included, FEMA director Julius Becton reportedly told him that "If I would not modify or remove the statement, he would have to either accept my resignation or he would terminate my duty" (House Committee on Energy and Commerce, 1986: 257-262). Petrone resigned, and FEMA removed the statement. Petrone, and Shoreham's opponents, maintained that FEMA had pressured him to either abandon his professional judgement or be fired, and that the ultimate source of the pressure was the NRC, the Department of Energy, and the "power lobby" (*New York Times*, April 15, 1986: II-5).

FEMA head Becton, on the other hand, said that Petrone's statement would have violated the agency's agreement with the Nuclear Regulatory Comission regarding the scope of the report:

> There have been published suggestions and many statements here that FEMA headquarters required the contents of that report to be altered. In one sense, this is correct. At my direction, the personal opinion of the Agency's former region II director [Petrone] was deleted from the final report. The pronouncement on his part is diametrically opposed to an Agency position he knew well. The...report itself was to make no finding or determination on the adequacy of the offsite preparedness. We were simply to discuss the results of the exercise. The insertion of those 29 words, which would have reached a conclusion as to the offsite preparedness, would have violated the FEMA-NRC agreement as well as our

testimony before the Congress (United States House Committee on Energy and Commerce, 1986: 278).

Representative Norman Lent also argued that the NRC's current regulations allowed for the possibility that utility plans could substitute for state and local action, and that Petrone was improperly inserting his personal views in an agency report (p. 267).

Thus, there were two views of the resignation. One was that Petrone had been pressured to resign for making a correct, important, and controversial statement. The other was that Petrone had no authority to insert a personal conclusion into the report; that the conclusion would have violated FEMA's agreements with the NRC and Congress; and that he was given a choice of following agency policy or being fired. In January 1987, the Acting Suffolk County Executive, Michael LoGrande, hired Petrone to coordinate the County's efforts to block the plant (*New York Times*, January 18, 1987: XXI-1).

Four days after Petrone's resignation FEMA released a generally favorable evaluation of the test, saying that "The team mobilized expeditiously and demonstrated the knowledge and capability to respond effectively to most scenario events." The Assessment noted some deficiencies requiring corrections, including delayed communications and inadequate performance by some dispatched in the mock evacuation. The Assessment also did not evaluate how the County and State boycott affected the exercise of the plan. Later, a LILCO representative said that "We are pleased with the high marks we got." A FEMA official, however, commented that "I'm not sure you can generalize that much," noting that certain deficiencies had been found and that it was normal to hold more than one drill for a nuclear plan to see if the applicant had remedied earlier deficiencies (*New York Times*, April 19, 1986: II-30; May 11, 1986: XXI-1).

The Federal Response to State and Local Opposition

On April 26, 1986, one week after FEMA released its initial evaluation of the LILCO exercise, a Soviet nuclear reactor at Chernobyl exploded and caught fire. The explosion blew the top off of the reactor, exposing the hot core to the environment. For three days radioactive smoke poured out of the reactor building, blowing over Europe and eventually touching the United States. Two hundred individuals, mainly plant workers and fire fighters, developed acute radiation sickness, and thirty one died in the weeks following the accident. Others would die subsequently. Forty nine thousand people were evacuated from around Chernobyl.

In the following days defenders of nuclear power noted repeatedly that reactors in the Soviet Union and the United States were designed very differently. In particular, Soviet reactors did not have the thick reactor walls and containment structures required in the United States to contain emissions, and the reactors' technologies differed greatly (Wilson, 1987). Nevertheless, the Cherynobyl accident energized the anti-nuclear movement, including those opposing Shoreham. In early 1985, according to one poll, respondents expressing an opinion split 26-70 when asked if they favored or opposed the building of more nuclear plants in the United States; shortly after Cherynobyl, they split 19-80 (Ahearne, 1987). Nora Bredes, the coordinator of the Shoreham Opponents Coalition, told a 150 person demonstration outside of the plant that because of the accident they should "rebuild the movement here on Long Island and stop Shoreham" (*New York Times*, May 11, 1986: XXI-1).

Opposition to Shoreham had spread across Long Island by the Fall of 1986. Virtually all candidates in the November 6 elections went on record as being against the plant, and candidates contended mainly over who opposed it first and most effectively (*New York Times*, October 16, 1986: II-7). The opposition's scope was important, because the greater the number of organizations committed to not cooperating with LILCO, the greater the difficulty LILCO would have in showing that it would be able to implement its emergency response plan.

For example, in June the Nassau County Board of Supervisors voted to refuse to allow LILCO to use the Nassau Veterans Memorial Coliseum as an emergency reception center. At the Atomic Safety Licensing Board's (ASLB's) hearings on the February emergency planning exercise, the Nassau chapter of the American Red Cross told the ASLB that it could not find schools or other private areas pledging to provide shelter to evacuees. The superintendent of the East Meadow School District in Nassau County told the ASLB that a majority of the county's 52 school districts would not agree to the use of school buildings or school buses by evacuees, saying that school officials disagreed with the LILCO plan in general and feared contamination of buses. FEMA's description of LILCO's exercise had reported that the Nassau chapter of the Red Cross "has made arrangements to use numerous facilities as congregate-care centers" in case of an emergency at Shoreham. A FEMA official now said that the Red Cross's inability to find facilities that would support the plan undercut that earlier statement; "Obviously, we have yet another problem there. We've counted on the Red Cross." Also, radio station WALK resigned as a primary broadcaster of emergency messages in case of an accident at the plant. Later, in August, 1987, the towns of Hempstead and North Hempstead declared that LILCO's decontamination trailers, an important part of the plan, violated local zoning ordinances (*New York Times*, September 28: 43; October 5, 1986: XXI: 8; August 16, 1987: XXI-10; *Newsday*, June 28, 1988). Thus, by 1987, it was clear that LILCO would receive no emergency planning assistance from the State or local governments; that, in fact, the governments and most participating not-for-profit organizations, pressed by an active coalition of anti-Shoreham groups, would actively try to block emergency planning for the plant.

Up to this time an applicant for an operating license had to demonstrate, through the emergency plan exercise, that they had a workable, effective plan. A valid exercise had to include state and/or local participation, because they would be critical actors in any actual emergency. To FEMA, the February 1986 exercise had tested *LILCO's* capabilities, and not the overall effectiveness of the Shoreham plan (Letter from Samuel Speck, FEMA to William Dircks, NRC, October 25, 1985). Because there would be no exercise for Shoreham involving state and local governments, presumably there would be no operating license. Senators D'Amato and Moynihan from New York, and Long Island's Congressional delegation in general, backed the State and County decisions.

The Long Island Lighting Company, and utilities in general, argued strongly that New York State and Suffolk County were misusing Federal emergency planning rules. Numerous members of Congress, the Department of Energy, the White House, and the *New York Times* and *Newsday* editorial staffs agreed with LILCO's position. In March 1987, the NRC proposed a rule undercutting the ability of state and local governments to veto a nuclear plant operating license by refusing to participate in emergency planning. The agency said that

> significant policy questions of equity and fairness are presented where a utility has substantially completed construction and committed substantial resources to a nuclear plant and then, after it is far too late realistically for the utility to reverse course, the state or local government opposes the plant by non-cooperation in off-site emergency planning. A forced abandonment of a completed nuclear plant for which billions for [sic] dollars have been invested also poses obvious serious financial consequences to the utility, ratepayers, and taxpayers. Finally, at least in situations where non-cooperation in offsite emergency planning is motivated by safety issues, vesting State or local governments with de facto veto authority over full-power operation is inconsistent with the fundamental thrust of the Atomic Energy Act whereby the Commission is given exclusive de jure authority to license nuclear power plants and to impose radiological safety requirements for their construction and operation (52 Federal Register 6981 [March 6, 1987]).

Accordingly, the NRC proposed a rule changing its policy on state and local participation at the design and testing phases of emergency planning. Under the proposed rule an applicant for an operating license would have to demonstrate that they had prepared an emergency response plan that would be effective *given* "reasonable State or local government cooperation." They would have to make a good faith effort to enlist state and local cooperation in emergency planning, would have to provide copies of the plan to the governments, and stand ready to cooperate with the governments in the future. However, even if state and local governments had not participated in the design or exercise of the plan, the NRC could approve the plan if it concluded that the plan would be effective *if the governments responded to an actual emergency.* Furthermore, under the new rule the NRC would presume that state and local government would respond in an actual emergency:

...if a plant began operation under the circumstances permitted by the proposed regulation change, and all administrative and judicial remedies available to plant opponents have been exhausted, it seems reasonable to expect that the governments involved more likely than not would change their position and cooperate in planning...The Commission believes that State and local governments which have not cooperated in planning will carry out their traditional public health and safety roles and would therefore respond to an accident. It is reasonable to expect that this response would follow a comprehensive utility plan (52 Federal Register 6982-83 [March 6, 1987]).

The NRC's general counsel called this presumption "the realism doctrine" (*New York Times*, February 27, 1987: IV: 18).

Commissioner James Asselstine dissented in the 4-1 vote approving the proposal. He said that *presuming* that state and local governments would cooperate rather than *assuring* they would cooperate, and *assuming* they would perform effectively rather than *testing* their performance in a simulated emergency, weakened emergency planning requirements. He also argued that the Commission's "assertions [on state and local participation in an actual emergency] are either irrelevant, insufficient or based simply on wishful thinking" (52 *Federal Register* 6985 [March 6, 1987]).

Similarly, FEMA commented that emergency plan exercises improved the coordination between utilities and government, and helped train emergency workers. Under the proposed rule those benefits would be seriously diminished, if not lost, because state and local officials might not have participated in the exercises. In those cases their responses in an actual emergency would be ad hoc and possibly ineffective. Finally, FEMA questioned the NRC's view that "State and local governments which have not been involved in emergency planning would nonetheless respond to an actual emergency and follow a comprehensive utility plan. FEMA has no data that would indicate what State and local government reactions might be in such circumstances." If the NRC accepted the rule, however, FEMA "would expect to continue our productive relationship [with the NRC]. We would also hope that the NRC would carefully monitor the participation of state and local governments in order to adjust requirements that might discourage such participation" (letter from Dave McLoughlin, FEMA, to Samuel Chilk, NRC, April 28, 1987).

Individuals and organizations opposing Shoreham, and the Seabrook plant in New Hampshire, argued that the rule change was an obvious attempt to remove barriers to operating licenses for the two plants over state and local objections. Mario Cuomo told an NRC hearing on the draft proposal that "You will not say the evacuation requirement should be dropped--you dare not. Then how can you say this proposal would fairly provide it? The plain fact is that this proposal would pretend to provide for safe evacuation, but in practical effect it would eliminate the requirement" (*New York Times*, February 25, 1987: 14. See also U.S. Senate Committee on Environment and Public Works, 1987a, 1987b).

On the other hand, other actors felt that the proposed rule change was a reasonable response to what they considered to be state and local misuse of the emergency planning requirements. A *New York Times* editorial said that "There's no question of usurping states' rights or police powers. All that the Commission staff proposes is that a plant should be licensed once a reasonable emergency plan has been drawn up, even if the local authorities say they won't cooperate...the [NRC] has already delayed too long in restoring its authority" (February 9, 1987: 18).

In fact, there was no evidence that effective emergency planning around Shoreham was not possible. In 1983, Suffolk County's consultants had not concluded that effective emergency planning was impossible, but only that the risks of Shoreham could not be reduced to zero. Because all risks could not be eliminated, Suffolk declared that the public safety could not be totally protected, and therefore withdrew from the planning. Consequently, LILCO's plan was critically flawed, but that was because of County and State non-cooperation and not because of the inherent problems of planning around Shoreham itself.

An Atomic Safety and Licensing Board, in calling the plan critically flawed in 1985, nevertheless highlighted this point:

> From the evidence of record the Board finds that no operating license shall be issued. We make this finding because the LILCO plan does not provide reasonable assurance that adequate protective measures can and will be taken in the event of a radiological emergency at Shoreham, as required....This determination is not based upon a finding that there is anything unique about the demography, topography, access routes, or jurisdictional boundaries in the area in which Shoreham is located. To the contrary, the record fails to reveal any basis to conclude that it would be impossible to fashion and implement an effective offsite emergency plan for the Shoreham plant...It should be stated at this point, that inasmuch as it is the LILCO plan, the inadequacies are ascribable to LILCO. However, to a significant degree the inadequacies resulted from and have been aggravated by Suffolk County's and New York State's opposition to the plan...Although we do not find their

opposition to be contrary to law, that action has helped to create a barrier to the implementing of an effective emergency plan (22 NRC 410 [1985], pp. 426-427).

In supporting the NRC's proposed rule change, LILCO maintained that

[I]t has now been demonstrated in NRC hearings that the entire 10-mile emergency planning zone (EPZ) around Shoreham can be evacuated more easily than the EPZ's around many other licensed nuclear plants. LILCO's analysis, accepted by the Licensing and Appeals Boards, shows a normal evacuation time of 4 hours and 55 minutes for the entire EPZ in normal weather and with traffic guides. These times compare favorably to other plants located in and supported by New York State. Comparable evacuation estimates for the Nine Mile Point nuclear site are 5 hours and 50 minutes, and for Indian Point, 7 hours and ten minutes. A comprehensive 1982 [NRC] study evaluating 52 nuclear plant sites on a consistent basis showed that half the plants studied had evacuation times in excess of that expected for Shoreham (cited in U.S. Senate Committee on Environment and Public Works, 1987a: 363-364).

LILCO also pointed out that the evacuation time estimates prepared by Suffolk County tended to be similar to LILCO's estimates whenever comparable conditions were modeled: "The much longer evacuation time estimates (15 to 18 hours) sometimes cited by Suffolk County require inclusion of extra-EPZ traffic and roads. They apply *only* to the EPZ-periphery travel times of voluntary evacuees who originate outside the EPZ from as far away as 40 miles from Shoreham, never enter the EPZ during their journey, and never come closer to Shoreham than the 10-mile EPZ boundary. They also require expansion of the EPZ to include a road now outside its boundary" (p. 364).

Assuming a ten-mile EPZ, LILCO's observation was correct. The longer periods cited by Suffolk referred to evacuation in the 10-20 mile zone not required by NRC regulations but demanded by Suffolk County (See Table 12 earlier, reproduced from Herr and Associates, 1983: 350). An Atomic Safety Licensing Board in 1988 commented similarly, regarding hospital evacuation time estimates, that "[Suffolk County and New York State's) credibility in this part of the proceeding was diminished by insistence that LILCO's evacuation times are unreliable or likely longer when the evidence showed that their own [Evacuation Time Estimates] tended to confirm rather than refute LILCO's" (28 *NRC* 355 [1988]).

On August 5, 1987, the House of Representatives rejected 261 to 160 an amendment to an appropriations bill that would have barred NRC approval of the rule change (*New York Times*, August 6, 1987: 16). On October 29, the NRC voted unanimously to adopt the rule essentially as proposed, stating that while state and local participation in emergency planning was "highly desirable and indeed is essential for maximum effectiveness of emergency planning and preparedness, Congress did not intend that the absence of such participation should preclude licensing of substantially completed nuclear power plants where there is a utility-prepared emergency plan that provides reasonable assurance of adequate protection to the public" (52 *Federal Register* 42078-79 [November 3, 1987]).

On February 17, 1988, the New York State Court of Appeals declined to rule on whether LILCO had the legal authority to implement an emergency response plan, concluding that the Federal government had regulatory jurisdiction over the plan (*Albany Times Union*, February 18, 1988). The NRC and FEMA scheduled another test of LILCO's emergency response plan for June, 1988. The NRC's rule change, if upheld in court, seriously diminished the legal ability of Suffolk County and New York State to block Shoreham by refusing to participate in emergency planning.

But by 1988 the political effects of the emergency planning dispute overshadowed LILCO's at least temporary legal win. Up to the early 1980s Suffolk County and New York State governments had supported or at least tolerated Shoreham. That changed beginning in 1982. By 1988 the position that the Shoreham area could not be safely evacuated had been made so often by so many State and local officials that the argument took on a life of its own, regardless of the original evidence for or against it. The Three Mile Island and then Chernobyl accidents were a powerful background to the argument. Political careers on Long Island generally depended on opposing Shoreham. Governor Mario Cuomo backed the County and made the abandonment of Shoreham a major priority of his administration. Cuomo's relentless opposition was critical. He more than any other individual could mobilize New York State's regulatory organizations, with which LILCO had to live, against the company. Furthermore, the Long Island Lighting Company, not a popular institution, probably would lose a battle for political support with a strong Governor. The Federal support for Shoreham seemed to be a relatively weak signal in this exceedingly hostile state and local environment.

In this setting the best that LILCO could hope for would be to escape the Shoreham problem as a viable, private utility, with or without the plant operating. The next chapter examines the events leading to a proposed settlement of the Shoreham controversy.

Appendix to Chapter 5
Chronology of Key Events

1973

April 12: Atomic Energy Commission announces it is granting construction permit for Shoreham.

October 26: Atomic Safety and Licensing Appeals Board affirms the issuance of construction permit for Shoreham.

1975

October: Suffolk County prepares the first emergency response plan for Shoreham in cooperation with LILCO.

1979

March 4: Three Mile Island accident occurs, resulting in a partial meltdown at the plant in Pennsylvania.

December: Following problems of emergency response after the Three Mile Island Accident, the Nuclear Regulatory Commission proposes new emergency preparedness rules.

December: Suffolk County Executive John Klein and LILCO sign agreement updating County's responsibilities in emergency planning.

1980

August: The Nuclear Regulatory Commission adopts final new emergency preparedness rules, requiring more extensive emergency planning for nuclear power plants.

1981

September: Suffolk County legislature approves a contract with LILCO to prepare an emergency plan for Shoreham.

1982

February: An attorney representing Suffolk County in the Shoreham operating license proceedings advises the County to cancel its emergency planning contract with LILCO. Suffolk advises LILCO that it is withdrawing from the contract.

May: LILCO forwards the initial Suffolk County-LILCO emergency plan for approval to New York State Disaster Preparedness Commission. Suffolk County protests, calling the Suffolk County Planning Department document "resource material" only.

May: Suffolk County announces that it will prepare its own emergency response plan for Shoreham.

June 5: Nuclear Regulatory Commission Atomic Safety Licensing Board begins Shoreham operating license hearings.

September: Pre-operational testing of the diesel generators at Shoreham begins.

September 30: Torrey Pines Technology issues a report on Shoreham, recommending changes costing approximately $1 million but finding plant's construction and quality assurance program to be satisfactory from the standpoint of safety.

December: Suffolk County releases its draft emergency response plan, calling for planning for evacuation of up to twenty miles around Shoreham, rather than the ten required by Federal regulations, based on what it calls special circumstances on Long Island.

1983

January: Suffolk County conducts hearings on proposed emergency response plan for Shoreham.

February 16: Suffolk County Executive Peter Cohalan says that it is not possible to have adequate emergency planning for Shoreham, concluding that Shoreham therefore should never operate.

February 17: Governor Mario Cuomo orders the State not to approve any emergency plan not approved by Suffolk County. Hours after the Governor's executive order, the Suffolk County legislature votes 15-1 to cease emergency planning for Shoreham on grounds that effective emergency planning for Shoreham is not possible.

February 23: Suffolk County asks Nuclear Regulatory Commission to halt Shoreham's operating license hearings on grounds that the County had withdrawn from emergency planning, making it impossible for the NRC to grant LILCO an operating license for the plant.

April 19: Governor Cuomo proposes creation of a "Fact Finding Panel" for Shoreham, to be chaired by State University of New York at Stony Brook President John Marburger.

May 12: Nuclear Regulatory Commission agrees to consider LILCO's emergency response plan, over Suffolk County's objections.

June 8: LILCO files a motion with the NRC for a low power license for Shoreham.

August 12: Crankshaft on one of Shoreham's emergency diesel generators breaks.

October 1: Marburger Fact Finding Panel on Shoreham meets in executive session to consider conclusions for its Final Report to the Governor.

October 5: Federal Emergency Management Agency accepts Nuclear Regulatory Commission's referral of LILCO's emergency plan.

October 24: Stone and Webster recommends that LILCO purchase 3 Colt diesel generators to replace the Delaval diesel generators.

November: LILCO orders diesel generators from Colt Industries.

December: Final Report of Marburger Panel is released. Panel concludes that Suffolk County's decision to withdraw from emergency planning is not so unreasonable as to warrant intervention by Governor Cuomo to overturn decision.

1984

January 10: LILCO announces it will withhold property tax payments to Suffolk County on grounds that County was overassessing the Shoreham plant.

January 30: Charles R. Pierce resigns as LILCO chairman, and is replaced by Dr. William J. Catacosinos.

May: Suffolk County Legislature's Budget Review Office concludes that property taxes would need to rise up to sixteen percent if Shoreham was not taxed as an operating unit.

June: Suffolk County faces budget shortfall of $200 million in its budget of $800 million, a budgetary shortfall aggravated by LILCO's witholding of property taxes for Shoreham.

October: Atomic Safety Licensing Board issues low-power testing license for Shoreham.

November: Suffolk County's bankers raise concerns about the effects of LILCO's withholding of property taxes.

November: Federal Energy Secretary Donald Hodel argues to Suffolk County Executive Peter Cohalan that, based on data from the Three Mile Island accident, it might be possible to reduce the planning area for evacuation around a nuclear plant from ten miles to two miles.

1985

May 31: Peter Cohalan orders Suffolk County to cooperate with LILCO in testing emergency response plan for Shoreham. LILCO announces that it will cease withholding property taxes for Shoreham, and will pay $131 million in back taxes to Suffolk County. Suffolk County Legislature and Governor Cuomo oppose Cohalan's executive order.

June 10: Responding to an appeal by the Suffolk County Legislature, a New York State judge nullifies Cohalan's executive order.

June 20: Nuclear Regulatory Commission asks the Federal Emergency Management Agency (FEMA) to conduct exercise of LILCO's emergency response plan.

August: Shoreham begins low-power testing.

November 13: Chairs of the Senate and House Committees in charge of Federal Emergency Management Agency's appropriations write to FEMA saying that it should not allow state and local opposition to nuclear plants to undermine emergency planning efforts.

November 14: FEMA confirms that it will conduct test of LILCO's emergency plan.

1986

January 11: Suffolk County Legislature passes a bill prohibiting exercise of emergency plan in the County without approval by the Legislature.

February 10: Federal judge bars Suffolk County from enforcing law preventing emergency planning exercise for Shoreham.

February 13: FEMA, the NRC, and LILCO conduct test of LILCO's emergency response plan.

April 14: FEMA Regional Director Frank Petrone resigns over wording of Post-Exercise Assessment. Petrone contends that the Assessment should specify that FEMA could not provide reasonable assurance that the plan would protect public safety because the local authorities did not participate in the exercise.

April 18: FEMA evaluates LILCO's test of emergency plan relatively favorably, but does identify note some deficiencies requiring corrections. Notes that emergency plan approval process normally requires more than one exercise.

April 26: Soviet nuclear reactor at Chernobyl explodes, spewing radioactive cloud over Europe which eventually reaches United States. Forty nine thousand people around Chernobyl are evacuated; thirty one die shortly, with more deaths expected from exposure to radiation.

June: Nassau County Board of Supervisors refuses to allow LILCO to use Veterans Memorial Coliseum as emergency reception center. Nassau Chapter of American Red Cross tells the Atomic Safety Licensing Board that it cannot find school districts or other areas willing to provide transportation or shelter to evacuees.

September: Radio station WALK resigns as primary broadcaster of emergency messages in LILCO's emergency response plan.

1987

March 6: After discussions in February, the Nuclear Regulatory Commission formally proposes that it be allowed to *presume* that state and local governments would assist in any actual nuclear emergency; and that it be allowed evaluate emergency response plans, including those submitted by utilities, with that presumption.

August 5: House of Representatives rejects by a vote of 267 to 160 an amendment to the Nuclear Regulatory Commission's appropriation which would bar NRC approval of revised emergency planning procedures.

August: Long Island towns of Hempstead and North Hempstead declare that LILCO's decontamination trailers, an important part of its emergency response plan, would violate local zoning ordinances.

October 29: Nuclear Regulatory Commission votes unanimously to adopt revision of emergency planning rules, allowing it to presume that state and local governments will assist in any actual emergency.

Chapter 6

TAKEOVER, SETTLEMENT, OR SHOREHAM?

One way for New York State and Suffolk County to attempt to block Shoreham was to refuse to participate in emergency planning. Another was to take over LILCO, one way or another, and then abandon the plant; or *threaten* a takeover to force the company to abandon it. Such maneuvering dominated the Shoreham controversy in 1988.

In the early 1980s Suffolk County began exploring the possibility of establishing a public power system in the County, replacing LILCO. At the time LILCO's electric rates were the second highest in the nation, and advocates of public power systems argued that such systems' electric rates tended to be lower than those of investor-owned utilities. Private systems ostensibly had an incentive to add to their rate base by expanding their capital facilities, even though it might be more efficient to import power from power systems with excess capacity. (This was a criticism which might have been appropriate at one point, but not in the 1970s or 1980s; see Navarro [1985].) Thus, they contended, the public New York Power Authority had access to less costly electricity than that generated by LILCO, both because of its low-cost hydropower and because its generating costs simply were less than LILCO's (Wilmott and Harrison, 1983: 61).

In June 1983, the consulting firm of Daverman and Associates reported to the County that acquiring LILCO's assets would cost Suffolk County about $1.4 billion, and that doing so would save Suffolk ratepayers between 4 and 30 percent on electric bills, depending on analytical assumptions. The Daverman report was not definitive on either the costs or benefits of establishing a public power system in Suffolk County. For example, the Suffolk County Budget Review Office noted that "If one thread was found to flow through our entire research and analysis [on the report], it was the absence of hard definitive dollar projections. Many of the conclusions reached by Daverman were based upon assumptions whose validity remains uncertain" (Cited by Long Island Lighting Company, 1984: 4).

Nevertheless, LILCO was a politically popular target because of Shoreham and the company's high electricity rates. In October 1984, ten members of the County Legislature introduced a resolution to establish a public electric utility in Suffolk County which would acquire LILCO's capital facilities and other necessary assets. The resolution would also prohibit the organization from owning and operating any nuclear power station (Suffolk County Resolution No. 2028-84, October 23, 1984).

The Establishment of the Long Island Power Authority

By late 1985 the Long Island Lighting Company was in an extremely weak position financially and politically. The price of LILCO's stock had declined from $17 in late 1982 to about $7 (rebounding from under $4 in mid-1984). It had no cash earnings, and so was borrowing to pay for short-term expenses and interest on previous borrowing. Banks repeatedly extended loan repayment deadlines to keep the utility operating. The company had paid no common stock dividends since March 1984, nor preferred stock dividends since October 1984. LILCO was severely criticized for its handling of a hurricane in late September which caused several days of power outages and $25-$30 million in damage to its transmission facilities on Long Island. Adding to the company's problem was the refusal of LILCO's Chairman William Catacosinos to return from a vacation to attend to the storm crisis (*New York Times*, October 1, 1985: IV-1).

Furthermore, in May 1986 the Suffolk County legislature voted to file a $5.4 billion suit against LILCO under the Federal Racketeer Influenced and Corrupt Organizations law, arguing that LILCO had pursued rate increases from the New York State Public Service Commission using arguments about Shoreham it knew to be false (*New York Times*, June 1, 1986: 37). As will be discussed in Chapter 7, this law suit became surprisingly and enormously important in late 1988 and 1989.

The Cuomo administration was fighting LILCO both on the need for Shoreham's electricity and LILCO's rates, in addition to the emergency planning issue. The New York State Power Pool, a consortium of utilities, had fined LILCO $1 million for failing to maintain an eighteen percent power reserve during a peak demand period in the Summer of 1985; the letter advising LILCO of the fine said that "It is realized that this deficiency is caused by LILCO's inability to operate Shoreham." The New York State Energy Office, however, argued that LILCO could meet current and expected power demands without Shoreham by reactivating mothballed facilities, conservation, and purchases of power from other utilities. The Energy Office report explicitly excluded Shoreham as a possible power source in its master plan for the period 1986-2000 (*New York Times*, December 29, 1985: XXI-9; April 20, 1986: XXI-1).

The Cuomo administration also argued that any rate increases the New York State Public Service Commission (PSC) might grant LILCO would only allow the company to pursue the Shoreham license. The Public Service Commission's members were appointed by the governor but, once confirmed, were independent. In February 1985 LILCO requested a 3.8 percent, $68.7 million rate increase from the PSC. In July the staff of the Department of Public Service (DPS) supported the increase during a rate hearing, maintaining that "the company's financial condition is such that the full level of revenue requested is the minimum necessary to put the company in a position to finance [its current debt and operating expenses] at reasonable terms." The DPS Staff also recommended that almost all of the increase go to meet LILCO's cash flow needs rather than provide for licensing and operation of Shoreham. In contrast, the State Consumer Protection Board argued that LILCO's original request presumed the operation of Shoreham, a plant that might never open. The Consumer Protection Board also said that any rate increase would allow LILCO to pursue a doomed project, that the company could afford a denial of the increase, and that, in any event, "The PSC [cannot and should not] take the position that LILCO should be kept out of bankruptcy at all costs" (*Newsday*, July 18: 1985).

In January 1986, the Public Service Commission granted LILCO the entire $68.7 million rate increase, concluding that the increase was necessary to improve the utility's dangerous financial position. The increase would add about $3.40 to the monthly bill of an average Long Island residential customer. The Governor, who had appointed four of the five current members of the PSC, argued that the company's financial troubles were due almost entirely to Shoreham, and that the rate increase would encourage LILCO to pursue a Shoreham license and thereby delay the search for acceptable power alternatives. The State Consumer Protection Board said that it would sue the PSC, with the Governor's support, to block the rate increase. Cuomo also commented that "If I could start all over with the Public Service Commission, without confirmation, would this be the same Public Service Commission? No." (*New York Times*, January 10, 1986: II-2).

After receiving the rate increase, the company and its banks negotiated a refinancing of $1 billion in debt. The company would pay $400 million of the debt immediately though sale of a $525 million bond offering and spread payment of the remaining debt from 1989 through 1992. The three-year grace period from 1986 through 1988 would presumably allow LILCO time to obtain an operating license for, and cash flow from, the Shoreham plant. It was generally recognized that without the refinancing LILCO may have declared bankruptcy in 1986. Cuomo's Executive Director of the Consumer Protection Board, Richard Kessel--who opposed any move increasing the chance of a Shoreham license--argued that "People should be paying for service, not to help extension of bank loans. Its sad, because it's the ratepayers who are going to be forced to pay for these loan extensions" (*Newsday*, January 8, 1986; *Wall Street Journal*, January 8, 1986).

By 1986, the leaders of the Democratic-controlled New York State Assembly and Democratic Governor Cuomo all had backed a public takeover of LILCO. Then, in January 1986, the Republican State senators from Long Island declared that "As a delegation, we unanimously favor public power for Long Island. We believe that a public agency would be able to best meet Long Islanders' energy needs at reasonable rates," adding that they had the support of the Republican Senate majority leader, Warren Anderson. Nineteeen eighty-six was an election year for the Governor and the entire New York State legislature, and proposing a public takeover of an unpopular electric utility was politically timely; 71 percent of respondents questioned in a *Newsday* poll favored the concept of public power on Long Island (*New York Times*, January 7, 1986: II-1; January 12: XXI-1; *Newsday*, November 26, 1985).

In June 1986 two panels evaluated such a takeover of LILCO favorably. A $200,000 study commissioned by Suffolk County concluded that a takeover was feasible and would reduce electric rates on Long Island (*New York Times*, June 6, 1986: II-2). Then, in late June, a State panel concluded that over a fifteen year period a public power authority (the "Long Island Power Authority", or LIPA) operating Shoreham would charge about 9 percent less for electricity than LILCO with Shoreham operating. Without Shoreham operating LIPA would charge about seven percent less than LILCO. The Report commented that "Potential ratepayer savings from LIPA would be realized because it would not pay Federal income taxes and would use a capital structure consisting entirely of tax-exempt debt. Consequently, LIPA's capital charges would be considerably lower than LILCO's charges from its projected capital structure and associated costs of capital" (New York State Long Island Public Power Panel, 1986: 11-12). However, the authors of the State study said that the estimated savings were sensitive to the utility's financial and operating conditions at the time the takeover would occur and other considerations. Reasonable changes in the study's assumptions might either eliminate or increase the savings (pp. 15-16).

One week after the State panel released its report the Governor, Senate (35-26), and Assembly (80-63) agreed on a bill establishing the Long Island Power Authority (LIPA). LIPA was empowered to evaluate whether a takeover would reduce electric rates below what they otherwise would be. If the takeover would reduce rates, it was authorized to negotiate a friendly takeover of LILCO through a stock purchase with LILCO. LIPA also was authorized to attempt a hostile takeover through stock purchases or condemnation of LILCO's assets. In addition, the law prohibited Shoreham from being included in LILCO's rate base unless it were "used and useful," precluding the possibility that LILCO would be allowed to recover the bulk of Shoreham's costs from ratepayers if Shoreham never operated commercially. The Long Island Power Authority's effective date was January 15, 1987. Announcing the agreement, Governor Cuomo explained that "LILCO rips off the people of Long Island. Now we are coming after you."

Several factors converged in the passing of the legislation, including the Shoreham dispute, the frustration with LILCO's rates and handling of the recent hurricane, and the electoral benefits of being aggressive against an unpopular utility in an election year. In addition, the studies by Suffolk County and the State Public Power Panel had concluded that the concept of public power--a "rallying cry for so many frustrations" with LILCO--arguably had economic merit (*New York Times*, July 3, 1986: 1; July 12: 31). But, originally, blocking Shoreham was the fundamental purpose of the prospective takeover. (In the future, proponents of the takeover did not want to stop there.) In October 1986, Cuomo told LILCO that it had until the January 15, 1987, start-up of LIPA to abandon Shoreham if it wished to avoid a takeover, and announced that he had asked Vincent Tese, the State Economic Development Director, to seek a settlement that assured "no Shoreham, and adequate, affordable energy." Later he added that "There is not a whole lot of time because in January the Long Island Power Authority falls in and I have to appoint people to the new authority. Once you do that it takes on a life of its own. If LILCO wants to wait for the Long Island Power Authority, terrific. If LILCO wants to make a deal, it better start moving." An aide to Vincent Tese noted that the governor "can appoint a board that will ratify an agreement worked out by the negotiating teams, or he can appoint a board that uses its option to condemn the company" (*New York Times*, October 23, 1986: II-5; November 2: XXI-1).

By late 1987 the Governor, by re-constituting the Public Service Commission, also was able to pressure LILCO through the rate setting process. Since the PSC had granted LILCO a $68.7 million rate increase in January 1986--an increase opposed strongly by the Governor--Cuomo had appointed five new members of the PSC, including Peter Bradford, the new Chairman. In 1987, LILCO requested a 5 per cent, $83 million rate increase on the basis of increased taxes, operating costs, and its share of a Consortium's investment in the Nine Mile II nuclear plant. The DPS Staff and the administrative law judge hearing the rate request concluded that a 3.1 percent, $50 million increase was warranted. But in November the Public Service Commission rejected any increase in a preliminary decision, stating that

> ...[T]he continued provision of cash flow enhancement has prolonged a financially debilitating deadlock over Shoreham. That deadlock and uncertainty have jeopardized the fundamental goal of electric utility regulation--the assurance of a reliable and reasonably priced power supply. The uncertainty raises costs, saps LILCO's financial resources, inhibits power supply planning, consumes the attention of top management, and poisons the company's relationship with its customers and the government that they elect. It is in no one's interest for this situation to continue (New York State Public Service Commission Opinion 87-26, final opinion issued December 3, 1987: 5).

Accordingly, the PSC told LILCO that it had to come before the Commission the following April to show that "it has developed and will implement reasonable plans to provide reliable electric service without generation from the Shoreham power plant." Vincent Tese, who had become Cuomo's chief negotiator with LILCO, said that the decision "should convince LILCO" that "there will be no more rate hikes as long as Shoreham casts a shadow over Long Island," and that "The current action by the P.S.C. is the dawn of a new era, and it might give LILCO cause to do a little introspection. The state and the Long Island Power Authority are ready, willing and able to sit down with LILCO and negotiate out some of the problems that we all have. LILCO is not going to talk until it is forced to talk. But as things get a little hotter, there will be increasing pressure on it to do so." The PSC formally rejected the rate increase in December (*New York Times*, November 6, 1987: II-1; November 9: II-1; *Newsday*, December 4, 1987). By December it was clear that the Governor had committed the full force of the State's executive branch and, through the appointments process, the independent Public Service Commission, against Shoreham. In that context LILCO began negotiations leading to the abandonment of the plant.

Takeover Or Negotiated Settlement?

Nine members were appointed to the board of the Long Island Power Authority (LIPA), including William Mack (Chairman), a Long Island developer; Richard Kessel, the executive director of the State Consumer Protection Board; Vincent Tese, the State director of Economic Development and chairman of the State Urban Development Corporation; Nora Bredes, the executive director of a coalition opposing Shoreham; John Adam Kanas, President of the North Fork Bank and Trust Company; long-time Shoreham opponents Leon Campo, Irving Like, and Stephen Liss; and Martin Bernstein, an attorney and Councilman from North Hempstead on Long Island.

The Long Island Power Authority was to evaluate whether a public power system on Long Island would lower electric rates below those charged by LILCO. If a takeover would reduce rates, LIPA was authorized--some LIPA members said mandated--to take over the company.

A takeover could occur in three ways. First, the State could condemn the utility's assets for public use by paying "just compensation." The critical issue would be the fair price for the assets. In 1986, Maurice Barbash of Citizens to Replace LILCO said that the fair price would be the market value of LILCO's outstanding stock, then $1 billion. At the same time a company spokesman said that the fair price would be the value of LILCO's assets less depreciation, then $18 billion (*Newsday*, November 25, 1985: 27; *New York Times*, January 8, 1986: II-2). Early in LIPA's life it was recognized that the State likely would not try to condemn LILCO's assets because of the method's inevitable litigation and delays, and the risk that the courts might reach a decision that would be highly favorable to LILCO and costly to the State.

A *friendly* or *hostile* stock purchase were alternative ways to take over the company. In a friendly takeover, LIPA would make an offer to purchase its stock to LILCO's Board of Directors. If the Board found the offer acceptable it would recommend the sale to the company's stockholders. The June 1986 Report of the Long Island Public Power Panel had recommended that the State make every effort to pursue a friendly takeover rather than a more risky condemnation proceeding or hostile takeover (New York State Long Island Public Power Panel, 1986: 2-3). If the LILCO Board would not accept the purchase offer, LIPA could attempt a hostile takeover. In that event LIPA would attempt to purchase enough shares directly from stockholders to acquire controlling interest in the company. The cost of either a friendly or hostile takeover would depend on the price of LILCO's stock and other factors. LILCO argued strongly that the State's conduct was unfair because it was positioning itself to purchase the company at an artificially depressed price by blocking Shoreham's opening, denying the company rate increases, and other maneuvers effectively driving down the price of the company's stock (A point made in a report done for LILCO by the National Economic Research Associates [1986: 14-15]).

The money for the takeover would come from the sale of tax-exempt bonds. In late 1987, a Federal tax bill included a proposal that withdrew tax-exempt status from bonds used to finance utility takeovers, but a provision exempting LIPA was eventually written into the bill (*Newsday*, December 18, 1987; *New York Times*, December 20, 1987: XXI-1).

One other option was for the State and LILCO to negotiate a settlement wherein the State would forego any takeover, grant LILCO financial help, and provide help with power supplies in exchange for LILCO's abandonment of Shoreham and perhaps other concessions. The choice of negotiated settlement, takeover, or continued conflict dominated the Shoreham controversy in 1988.

Negotiations

The State and LILCO began negotiations over the future of the company and Shoreham early in 1987, at the same time as LIPA began to evaluate if and how to do the takeover. In the early negotiations the State proposed that an oversight board, other than the Public Service Commission, monitor LILCO and set its rates, and that the State give LILCO financial and generating help in exchange for LILCO's abandonment of Shoreham. Alternatively, LILCO proposed that the State buy all of LILCO's power plants and assume the responsibility for generating power on Long Island, and allow LILCO to reorganize as a nonregulated holding company which could enter profit-making lines of business. Those negotiations broke off in Spring 1987 (*Newsday*, December 17, 1987).

By late 1987 circumstances encouraged both the State and the Long Island Lighting Company to return to negotiations. At the time New York State ran an increased risk that the Federal Nuclear Regulatory Commission would license Shoreham. In October 1987 the NRC had approved the change in emergency planning requirements which potentially eliminated state and local governments' emergency planning veto of operating licenses for nuclear plants. On the other hand, LILCO also faced larger financial and political risks in continuing the effort to have Shoreham licensed. In denying LILCO a rate increase the New York State Public Service Commission had implied strongly that LILCO would receive no more rate increases until Shoreham was settled. Also, the Long Island Power Authority was moving toward the moment when it might actually attempt to take over the company (*New York Times*, May 29, 1988: 30; *Newsday*, May 29, 1988).

Although the State's negotiations formally were separate from the Long Island Power Authority's proceedings, two members of LIPA were the State's principal representatives--Vincent Tese, the State Director of Economic Development, and Richard Kessel, the Executive Director of the State Consumer Protection Board. LILCO's central negotiator was its Chairman William Catacosinos. Other officials, bankers, and attorneys were involved on both sides. Tese said that he viewed the negotiations beginning in November as a "last stab" at a settlement before a public takeover.

As the negotiations continued into January, Kessel said that "Obviously we are going to do no deal that involves the operation of Shoreham today or forever. And obviously they [LILCO] are looking for some kind of agreement that leaves the company whole as an investor-owned utility." LILCO Chairman Catacosinos said that the negotiations were proceeding "as rapidly as one could hope for, given the complexity of the issues that we have to deal with," but that reaching a settlement could take thirty to sixty days. A utility stock analyst said that "My guess is that they will work out some sort of compromise involving not operating Shoreham forever or at least putting it in mothballs. From talking with the company on and off I've found that they are not wed anymore toward running Shoreham." Mario Cuomo continued to signal support for public power on Long Island and to castigate LILCO, although it was clear that he was open to the possibility of a settlement (*New York Times*, January 24, 1988: XXI-1; January 25: II-3; *Newsday*, January 8, 1988; February 6).

However, the negotiations disturbed many other opponents of Shoreham, including some LIPA Board members hoping to begin a public takeover of the company. They feared that Tese, Kessel, and Chairman William Mack were stretching out LIPA's deliberations to allow time for a negotiated settlement. In January members Leon Campo, Nora Bredes, Irving Like, and Stephen Liss wrote a letter to Mack charging that "We cannot accept any further delays in our work for the sake of these negotiations. There is no evidence that LILCO has negotiated in good faith and we do not anticipate a change in its attitude or action." Leon Campo said that "There`s a deliberate intent to stall LIPA at the midnight hour for an indefinite period of time to accomodate LILCO in these negotiations...and it comes from the highest places in the state. LIPA is on a very, very deliberate hold pattern and it's because of the negotiations." Tese called the letter and the charges "political hackery" (*New York Times*, January 24, 1988: XXI-1; *Newsday*, February 27, 1988).

Campo and others objected mainly because a negotiated settlement, although leading to an abandonment of Shoreham, would leave LILCO as an investor-owned utility. Citizens to Replace LILCO, an organizational leader of the public power movement, issued radio advertisements saying "This is a warning. Long Island's victory over Shoreham and LILCO is in danger. Citizens to Replace LILCO has learned that some political appointees are negotiating secretly to bail out LILCO with your money," telling listeners to complain to the Governor. Maurice Barbash, the chairman of the organization, accused Tese of "negotiating with the enemy while the people sit in the dark stewing," and that the organization would spend as much as necessary to guarantee "which horse the governor rides to victory...a takeover, rather than a negotiated settlement."

Also, legislators who had sponsored the law creating LIPA announced they would hold hearings because the law "gave [LIPA] a specific mission to bring the virtues of public power to Long Island and it appears the board may have lost track...and instead is negotiating the closing of Shoreham in exchange for dropping the public power initiative." Finally, Irving Like argued at one meeting of the Long Island Power Authority that "If LILCO remains in existence its hegemony will continue. All of its connections and all of its media hype, all of that dynamic will continue and will seek to undermine a settlement and open Shoreham." Like also argued that LIPA was obligated statutorily to proceed with a public takeover if it found that a public power system would have lower rates than LILCO's. In contrast, Tese and Kessel maintained that in such an event LIPA was *authorized* but not *required* to attempt a takeover--that it could decide that a negotiated settlement would be a more desirable outcome than a takeover (*New York Times*, February 29, 1988: II-3; March 6: XXI-1; *Newsday*, February 27, 1988; March 1).

In early March LIPA members agreed to set a deadline on negotiations, beyond which the Authority would need to begin work on the takeover. Tese asked LILCO to postpone its annual stockholders meeting, then scheduled for May 17, for one month to allow more time for negotiations. The extension was necessary because LIPA would need about sixty days to prepare for a possible proxy fight at the meeting as a way to take over the company. Thus, unless the meeting date were extended, LIPA would have to decide to proceed with the takeover and begin the proxy action by about March 17, presumably precluding further negotiations. However, LILCO refused to delay the stockholders' meeting. Tese and others favoring a negotiated settlement lambasted LILCO for the refusal and questioned whether the company was serious about the negotiations, Tese saying that "obviously we need the 30 extra days to get this done. But let's put it this way, if Catacosinos and whoever he's got there want to lock themselves in a room for the next three days, then maybe we can get something done...Other than that, we're extremely disappointed and LIPA's going to have to move forward" (*Newsday*, March 3; *New York Times*, March 4: II-2).

The negotiations were to resume the next week, although at the same time LILCO formally asked LIPA to make its best offer for the purchase of the company. Catacosinos said that he offered to continue negotiations even while asking for a LIPA offer because he wanted to "display our willingness to seek a solution to the crisis whether it be by settlement or sale" (*New York Times*, March 5, 1988: 29; *Newsday*, March 5).

In March, Lazard Freres and Company gave LIPA its final "Comparative Rate Analysis of Public Acquisition of the Long Island Lighting Company." The Analysis concluded that by taking over LILCO at a common stock price at $8.62 a share, LIPA would be able to charge electricity rates 12 percent lower than LILCO's. The main reason was that LIPA would would be able to secure financing at lower interest rates because its bonds would be tax exempt (Lazard Freres and Company, 1988a). With the Lazard Freres analysis in hand and the deadline for the proxy fight approaching, LIPA's decision whether or not to attempt a takeover likely would come that month.

Vincent Tese argued against any takeover on the grounds that LILCO clearly was headed for bankruptcy, and purchasing the stock was the equivalent of bailing out the company and its shareholders; in all likelihood Tese also wanted to put off a takeover attempt to allow the negotiations to continue. On the other hand, Like, Campo, Bredes, and Liss argued that a takeover would be less of a bailout for LILCO than a settlement, and demanded that it commence. As LIPA members lobbied each other for votes in the coming month, takeover proponents recognized that they would need the support of Chairman Mack, Tese, and banker John Kanas. The three had the strongest connections and legitimacy with industry, and without their support LIPA would have a difficult time raising the money for the takeover regardless of the majority vote.

After the internal lobbying, and a six hour meeting on March 28, the Long Island Power Authority unanimously approved an offer to purchase LILCO. The offer was unattractive to common stock holders but, interestingly, was attractive to preferred shareholders. Holders of common stock would receive $8.75 per share, the average trading price over the previous thirty days but below current market value. On the other hand, holders of preferred stock would be paid the stock's redemption prices plus all accrued dividends. LILCO had not paid holders of preferred stock any dividends since 1984, and under LILCO's by-laws the preferred shareholders could elect a majority of the company's Board of Directors--including a Board sympathetic to the LIPA offer--if they had not been paid dividends for eight quarters.

The takeover vote was unanimous even though the LIPA Board was divided politically. This was probably so because the offer appealed in certain respects to both factions. Those who wanted the takeover to succeed arguably saw the offer as appealing to the preferred shareholders, who under the circumstances had the power to elect a Board of Director's majority sympathetic to LIPA. On the other hand, those opposing a takeover might have supported the offer as being unrealistically low for the common stockholders--and thus as likely to fail--but as something to intensify the pressures on LILCO management to negotiate a settlement. Despite the offer, LILCO said that it was willing to continue the negotiations, and Richard Kessel commented that "I think it would be foolish for us to say that there should be no more negotiations between the state and LILCO" (*Newsday*, March 22; March 23; March 31; April 1, 1988; *New York Times*, April 2: 26; April 3: XXI-1; April 4: II-3; Charles Mack letter to LILCO Board of Directors, March 29, 1988).

On March 31 Catacosinos, Tese, and investment bankers met at Tese's Manhattan town house in the pivotal session of the talks. At that point there had been a positive takeover vote. Also by that point, Tese and Catacosinos, along with their two bankers, were negotiating without others present. For whatever reason, over the two hour period Tese and Catacosinos were said to agree on virtually every major issue of the conflict. Tese recalled Catacosinos as saying

> "Vince, I want a transaction. I know you want a transaction. We can start right now." I just got the feeling after that meeting that Bill [Catacosinos] had crossed the line, that he'd crossed the line and decided 'I'm going after an agreement.' We probably accomplished more in that meeting than we accomplished in three years. [The meeting] cut through...a lot of the issues that would be on everybody's dream sheet but that in the real world are just not practical...We eliminated all of them except one, which I've always called a baloney issue [the handling of Suffolk County's civil suit under the RICO law against LILCO].

After the meeting Tese told Mario Cuomo (whom Catacosinos had come to label "Conan the Destroyer") that he was confident that a deal could be made (*New York Times*, May 29: 30; *Newsday*, May 29, 1988).

In mid-April LILCO called the Long Island Power Authority's offer to purchase the company a "firesale" price for its stock, rejected it, and did not make a counter-offer. LIPA Chairman Mack said that LILCO's failure to bargain showed that it was not interested in discussing a friendly takeover. But on April 25, the Authority voted 6-1 (with Bernstein opposing and Tese abstaining) to increase the offer for common stock from $8.75 to $10. Analysts still called the offer too low to succeed. Prior to increasing the offer LIPA filed documents with the United States Securities and Exchange Commission indicating that it also was preparing a proxy fight for a hostile takeover, and on April 29 voted 5-2 to begin soliciting proxy votes from shareholders (Bernstein and Kanas opposing, Tese abstaining. Leon Campo was absent, but would have voted for the takeover). Thus, LIPA had a "friendly" offer to LILCO outstanding while simultaneously undertaking a hostile takeover. On May 5 LIPA increased the "friendly" offer to $12 per share (*Newsday*, April 16, 1988; April 19; April 21; April 26; April 30; May 6).

By late April 1988 it also appeared that Tese and the company had come close to a negotiated settlement, with Mario Cuomo reviewing the outline. On April 27, Tese said that only "loose ends" stood in the way of a settlement under which electricity rates would "compare favorably to those under public power" (*Newsday*, April 28). The preliminary settlement's details emerged over the next two weeks as Tese, Kessel, and the Governor tried to line up support for it on Long Island and in the State legislature. LILCO would sell Shoreham for $1 to LIPA or the New York Power

Authority, which would then decommission the plant at LILCO's expense. LILCO would receive rate increases of about five percent per year for ten years, with the first three increases guaranteed and the following seven "targeted increases." The New York State Power Authority also would build several new power plants on Long Island to cover new electricity demands, eliminating LILCO's need to build new plants of its own until after the year 2000. Governor Cuomo, in publicly backing the idea of a settlement on May 16, indicated that talks continued and no final deal had been made.

On May 22 Cuomo said that LILCO had to agree to a final settlement within "the next few to several days." On May 25 he said that the settlement had to be reached by midnight or the State would break off talks. The last obstacle was LILCO's insistence that the State convince Suffolk County to withdraw its lawsuit of LILCO under the Federal Racketeer Influenced and Corrupt Organizations law, a demand the State called unfair after labelling the pending lawsuit "baloney." Shortly after the midnight deadline passed Cuomo said that the talks were off. At that point Catacosinos agreed to the final settlement.

The Settlement

Under the final terms LILCO would sell Shoreham to the Long Island Power Authority for $1. LIPA would then contract with the New York Power Authority to decommission the plant. It was expected that decommissioning would cost about $400 million; LILCO's ratepayers would pay for decommissioning. LILCO agreed to not operate Shoreham beyond the currently authorized five percent level for 90 days after the settlement had been filed with the Public Service Commission, which had to approve it.

The settlement established a "Financial Resource Asset" of about $3.7 billion ($3.656 billion) into LILCO's rates, which was equivalent to allowing most of Shoreham's "prudently incurred" costs into its rate base. LILCO would recover these costs over a forty year period. Under the settlement the Public Service Commission would grant LILCO rate increases of five percent per year for the following three years, and would target increases of five percent in the following seven years. Rates of return for the next several years were specified; if for whatever reason LILCO's earnings exceeded a designated cap the excess would be returned to ratepayers. The State would issue $100 million in tax-exempt industrial development bonds each year for the next five years to help LILCO restructure its debt. LILCO also would take a $2.5 billion Federal tax write-off for the Shoreham plant. The intention of the agreement was to enable LILCO to pay all back dividends on preferred stock by the first quarter of 1989, to resume payment on common stock dividends by that time as well, and to restore LILCO's debt, currently considered quite risky, to an investment-grade rating. Investment-grade securities are considered less risky and require lower interest rates, and so improving LILCO's bond rating would allow the company to finance its activities at lower interest rates.

The New York Power Authority would assume the responsibility to meet new demands for electric power on Long Island by building two 240 megawatt gas turbines for peak periods of electricity use, building two and possibly three 300 megawatt plants to meet basic power needs, and providing transmission and other assistance, all paid for through LILCO's rates.

LILCO would restructure its management, and make its "best efforts" to ask its Board of Directors to place a member of the Long Island Power Authority on the Board. The settlement terminated several court cases involving the State and LILCO, although one case it did not affect was Suffolk County's suit of LILCO under the Federal Racketeer Influenced and Corrupt Organizations law. Finally, the Long Island Power Authority would continue to monitor LILCO and retain the authority to attempt another takeover should it decide to do so. LIPA also would be consulted on decisions concerning future sites for power plants, and would work on conservation programs with LILCO.

The agreement was to be signed by Catacosinos, Cuomo, Tese, LIPA Chairman William Mack, Consumer Protection Board head Richard Kessel, New York Power Authority Head Richard Flynn, State Energy Commissioner William Cotter, Department of Public Service Staff representative Richard King, the Presiding Supervisor of the Town of Hempstead and the head of the Nassau County Board of Supervisors Joseph Mondello, Nassau County Executive Thomas Gulotta, the representative of the Suffolk County Legislature, Suffolk County Executive Patrick Halpin, and the New York Attorney General. Some observers said that Cuomo included the officials from Suffolk and Nassau counties in order to spread the responsibility for the substantial rate increases to follow. Cuomo said their names were listed in order to spread the credit for the deal (See Settlement Agreement-LILCO Issues, June 16, 1988; Memorandum of Understanding Concerning Proposed Agreements on Power Supply for Long Island [undated]; Asset Transfer Agreement Between Long Island Lighting Company and Long Island Power Authority, June 15, 1988; *Newsday*, June 17, 1988).

Under the settlement LILCO would receive enough financial assistance to remain a solvent investor-owned utility, and Shoreham would be dismantled. A utilities analyst commented that "although to the extent that ratepayers are losers because they're paying more, they may be winners in that this is an insurance policy if they're that worried about safety. Greater safety and peace of mind" (*Newsday*, June 19, 1988). Cuomo, Tese, Kessel, *Newsday*, and numerous other actors backed the deal as the best that could be had under the circumstances. Those who opposed Shoreham and backed the deal argued that if it were not accepted LILCO would be free to pursue an operating license from the Nuclear Regulatory Commission, and that given the NRC's amendment of its emergency planning requirements there was a good chance that LILCO would win the license.

The agreement also drew opponents from very different camps, particularly when its cost became clearer. A core of strong public power advocates opposed it immediately because LILCO would remain an investor-owned utility and because of the ensuing rate increases. In contrast, other actors rejected the idea of abandoning Shoreham because of the rate increases and a fear that the same type of opposition that brought down Shoreham would, on a smaller scale, block alternative power plants.

On June 22 LIPA voted 6-0 to accept the agreement. At the meeting Lazard Freres and Company, LIPA's financial consultant, presented an analysis which concluded that electricity rates would increase 4.6 percent per year over the next ten years under a public takeover, compared to 4.9 percent under the settlement and 5.2 percent with Shoreham operating. The analysis noted that a public takeover would cost ratepayers somewhat less than the settlement because a public authority would be exempt from Federal income taxes. Both scenarios would cost ratepayers less than Shoreham's operation because under both LILCO would phase out its large property tax bill for the plant (although there would be compensatory payments to the local governments to ease the loss of the taxes), because of Federal tax savings, and because both avoided heavy recurring capital expenditures that, Lazard Freres said, the NRC would likely require for Shoreham in the coming years (Lazard Freres and Company, 1988b, 1988c; see Table 13).

Public power advocate Leon Campo voted for the settlement on the grounds that the guarantee that Shoreham would not operate was worth the cost of the settlement over the takeover. However, Irving Like called it "an overly generous bailout of a grossly mismanaged, arrogant company." Like and Nora Bredes argued that the settlement was being "railroaded" through with little time allowed for study and debate, and left the meeting in protest before the vote; Steven Liss abstained. Also at the meeting, LIPA voted to place Campo on LILCO's Board of Directors as its representative (Like received two votes) (*Newsday*, June 23, 1988).

Projected Costs of the Settlement and Alternatives

In the following month analysts disagreed over the relative costs of electricity under LILCO with Shoreham operating, the settlement with

Table 13. Lazard Freres Comparison of Revenue Requirements Under Alternative Scenarios (Present Value in Millions, 1989-2003)

	LILCO With Shoreham	LILCO Under Settlement	LIPA
Operations and Maintenance	$10,284	$10,358	$10,358
Taxes Other Than Federal Inc.	3,848	3,205	3,176
Federal Income Tax	1,132	1,100	0
Capital	7,271	7,453	8,143
Gas Savings	--	0-7	23
Total Cost of Service	$22,536	$22,117-22,124	$21,701

Note: Components do not add to totals because of rounding.

Source: Lazard Freres and Company, 1988b, 1988c.

LILCO operating, and a takeover of LILCO by the Long Island Power Authority. Lazard Freres had projected that a Long Island Power Authority takeover offered a 1.8 percent savings (about $400 million) over the settlement. But the State Department of Public Service pointed out that "more than half of the claimed savings is attributable to avoidance of federal taxes," and that avoided taxes are not true savings because "other taxpayers must make them up or diminished services will result." Similarly, LIPA's projected borrowing costs were relatively low because its debt would be tax exempt, but again the "savings" would have to be made up by taxpayers and thus were not true savings. Finally, the LIPA scenario presented many uncertainties. There was no guarantee of the extent to which LIPA's financing would be exempt from Federal taxes. There also was no assurance that in the inevitably lengthy process of a takeover

LILCO would not obtain an operating license for Shoreham. The Department of Public Service noted that "the ostensible...savings [of a takeover] would be more than fully offset by either of these factors" (New York State Department of Public Service, 1988a: 41-49).

Similarly, the State Consumer Protection Board's analysis concluded that electricity rates generally would be lower under the settlement than under either a LIPA takeover or Shoreham operation (See Table 14). The settlement would be less costly than Shoreham's operation because under the settlement a non-nuclear LILCO would a less risky investment and therefore able to acquire capital at lower costs. This was so particularly if the settlement were approved--as it would have to be--by the State government, the utility, and the State Public Service Commission. Additionally, LILCO would avoid substantial capital improvements that, the Consumer Protection Board argued, the NRC probably would require at Shoreham in the near future.

Echoing the Department of Public Service's position, the Consumer Protection Board argued that the settlement also was superior to a takeover because of uncertainty regarding Federal tax treatment of LIPA under a takeover, doubt over whether the takeover would be successful, the risk that Shoreham would be licensed during litigation over the takeover, and other factors. "Placed in the proper perspective," Richard Kessel said, "any arguable difference between rates under the settlement and a LIPA takeover is insignificant given the certainty with which the Settlement will achieve the basic policy objectives of closing Shoreham and ensuring safe and affordable energy" (New York State Consumer Protection Board, 1988: 27-29).

But other analysts criticized the settlement strongly as being inferior to a LIPA takeover of LILCO. Union Associates, a consultant for the Suffolk County Legislature, pointed out that under the settlement LILCO would recover all but $377 million of Shoreham's prudently incurred costs through the settlement's Financial Recovery Asset, that it would write-off the remaining $377 million, and thus "the amount charged to ratepayers will equal no less than the sum LILCO would receive if Shoreham were to operate. In other words, ratepayers will pay 100% of Shoreham's so-called 'prudent' cost, the same as if the plant operated." Union Associates also argued that the Public Service Commission's analysis of the settlement's costs "does not go past the eleventh year (unlike the LIPA projections which extend for 15 years into the future). This is cause for concern because the Settlement is structured to delay major costs until the eleventh year and beyond" (emphasis in original), including the costs of decommissioning Shoreham. Finally, Union Associates maintained that the State government's analysis underestimated LILCO's cost of capital and overestimated LIPA's, and that overall LIPA enjoyed a long term rate

Table 14. Consumer Protection Board's Estimates of Base Rates (Cents Per Kilowatt Hour)

Year	Settlement	LIPA @ $16 Per Share*	LIPA @ $12 Per Share*	Shoreham Operation**
1989	11.93	12.11	12.10	13.21
1990	12.52	12.61	12.61	13.88
1991	13.11	13.22	13.21	14.57
1992	13.73	13.83	13.81	15.30
1993	14.39	14.44	14.42	15.98
1994	15.07	15.05	15.02	16.71
1995	15.79	15.76	15.73	17.47
1996	16.54	16.46	16.43	18.25
1997	17.32	17.18	17.14	19.07
1998	18.14	18.00	17.94	19.92
1999	17.27	18.81	18.75	20.83
2000	17.39	19.73	19.66	21.77

* Projected per share cost of LILCO's stock under a takeover.

** Projected from Public Service Commission Opinion No. 86-14

Source: New York State Consumer Protection Board, 1988: 25.

advantage of 20 percent over LILCO under the settlement (Union Associates, 1988).

Energy Systems Incorporated, commenting for Suffolk County, claimed that both Lazard Freres and the State government understated the settlement's costs and overstated the takeover's. Neil Talbot of Energy Systems said that Lazard Freres and the State underestimated LILCO's costs of capital under the settlement. Talbot also said that Lazard Freres overestimated the takeover's costs because it posited an excessively high price for LILCO's common stock ($19 compared to what the consultant called a more realistic $15) and inserted what Talbot called an unjustified $600 million "contingency" item in the projected costs. After changing these and other assumptions Energy Systems estimated that the takeover, compared to the settlement, would reduce revenue requirements by $2.3

Table 15. Energy Systems, Inc. Estimates of Base Rates (Cents Per Kilowatt Hour)

Year	Settlement	LIPA Operation ($15 Per Share)	Typical Annual Savings ($'s)*
1989	12.02	11.24	$115.4
1990	12.65	11.85	114.3
1991	13.32	12.40	111.3
1992	14.02	13.06	105.8
1993	14.75	13.66	110.4
1994	15.53	14.70	82.7
1995	16.34	15.36	78.3
1996	17.20	16.07	84.5
1997	18.10	16.66	96.2
1998	19.05	17.14	103.4
1999	20.05	17.88	113.4
2000	21.11	18.38	118.6
2001	22.21	18.95	134.7
2002	23.38	19.54	138.5
2003	24.26	20.14	$165.4

* Energy System's estimate of annual savings for a residential customer's electric bill for 500 kilowatt hours per month.

Source: Energy Systems, Inc.,1988: 19-20.

billion over a fifteen year period and appreciably lower electricity rates (Energy Systems, 1988; see Table 15).

The Staff of the Department of Public Service and the Consumer Protection Board both rejected Talbot's conclusions. A DPS Staff member argued that Talbot's study "selectively picks and chooses assumptions that unreasonably favor a LIPA scenario over the Settlement. If these errors and omissions are corrected, the Settlement is superior to the takeover" (New York State Department of Public Service, 1988b: 27). In preparing his analysis Talbot used assumptions made by the March 1988 Lazard Freres study which evaluated the prospective LIPA takeover of the Long Island Lighting Company. However, Lazard Freres had issued a subsequent report in June including updated information and the settlement's final terms; the DPS and CPB staff argued that Talbot should have used this updated information. In fact, they argued that Talbot selectively used the assumptions of the March and June Lazard studies in ways "which...would make LIPA look better vis-a-vis the settlement" (p. 28). The Department of Public Service and the Consumer Protection Board also criticized Talbot's assumed $15 per share takeover price, projections of interest rates, elimination of the $600 million LIPA contingency fund ("which LIPA's own consultant does not support"), and other decisions allegedly biasing the case against the settlement and in favor of a LIPA takeover.

The studies discussed above were presented at the hearings on the settlement conducted in the Summer of 1988. The hearings were conducted before Administrative Law Judge Frank Robinson, who would recommend, but not order, a decision on whether or not the Public Service Commission should accept the settlement. The debate essentially was over the settlement versus the takeover; Shoreham's operation was not considered a realistic option.

But on September 1 Robinson concluded that in his view Shoreham ought to be opened and operated. Robinson wrote that his analysis of the record

> suggests that the settlement is not in the public interest because it would scrap a completed $5.3 billion power plant whose operation should save Long Island money, would avoid a real danger of power supply deficiencies, and would be safe, clean, and environmentally beneficial; and that the settlement's ostensible benefits are not a sufficient tradeoff, because the power supply arrangements would be moot if Shoreham operates, and its rate arrangements can be (and would have to be) implemented independently. If, however, the Commission is willing to sign Shoreham's death warrant, the settlement is not otherwise objectionable; its rate arrangements are reasonable, and there would be no sound alternative scenarios (Robinson, 1988: 4).

Robinson rejected the argument regarding the difficulty of evacuating Long Island in the event of an accident on two grounds. First, NRC regulations called for possible evacuation of an area ten miles around the plant, not the entire Island. Second, he wrote that the evacuation issue was "overblown given the virtually infinitesimal chance of an accident actually requiring evacuation. There has never been such an accident in the thousand-plus reactor-years of commercial nuclear operations in this country. To scrap a $5.3 billion asset based on such an alleged safety concern is irrational in light of the multitude of far likelier threats and dangers that people live with in equanimity." Chernobyl "was a very different kind of plant which would never have met the vastly stricter U.S. safety standards." Three Mile

Island "also scared a lot of people, although no tangible public harm occurred despite incredible bungling by the plant operators--demonstrating how safe U.S. reactors really are. And that episode, of course, inspired a massive series of additional safety requirements."

Robinson argued that the risks associated with power outages on Long Island were far greater causes for concern than the risks of an accident, and that Shoreham "likely would solve Long Island's power supply problem for at least the next decade" (pp. 59-63). In contrast, the settlement assumed that the required additional peak load plants would be installed in time to meet power demands in 1989; that "ambitious (indeed, unprecedented)" conservation targets were met; that a Long Island Sound cable to transmit power to the Island would be completed on time; that cogeneration would emerge as a major force on the Island; and that "conservative" peak load forecasts would not be exceeded. "The safety margins," he said, "are skimpy, with little leeway for deviations between what is projected and what actually occurs. Thus, Shoreham's abandonment would leave Long Island vulnerable to power supply deficiencies."

"Even if everything contemplated by the settlement does fall compliantly into place," he added, "and Long Island's power needs are met without Shoreham, a price will have to be paid for this." Nationally, LILCO was the utility most dependent on an expensive and potentially unreliable supply of foreign oil, and the settlement would continue this situation. The environmental effects of burning fossil fuels--respiratory ailments, acid rain, the "greenhouse effect," etc.--would continue, "an impact whose magnitude vastly exceeds the health effects of nuclear reactors... Given these real and grave concerns, to tear down a clean, operable non-fossil power plant seems tragic" (pp. 65-66)

He also questioned the conclusion that either the settlement or a LIPA takeover would be cheaper for ratepayers than Shoreham's operation. The primary data for that conclusion were presented in the Lazard Freres study done for LIPA, a study generally supported by the Consumer Protection Board, the Department of Public Service Staff, and LILCO. Robinson maintained argued that "The study does not show...that to dismantle Shoreham is economically preferable to operating the plant. In fact, it shows the opposite." Under the settlement LILCO had agreed to forego from its rate base $244 million of Shoreham's costs. However, such an agreement was independent of the abandonment of the plant, and thus was not a unique benefit of either the settlement or a LIPA takeover. For example, LILCO could agree to forego recovery of the same $244 million in an arrangement under which the plant operated. "This correction," he said, "almost neutralizes the rate disadvantage shown for the Shoreham-in scenario over the first 15 years." Also, in the Lazard Freres study "the no-Shoreham scenario benefits from lower assumed property taxes. This ostensible rate savings is illusory because the missing property taxes would have to be made up by Long Islanders. Again, though, it is possible to eliminate this distortive factor and thereby come closer to isolating the true economics of Shoreham's operation. With this correction, the 0.3% rate disadvantage (over the first 15 years) to Shoreham's operation becomes a 0.1% advantage. When combined with the previously discussed adjustment concerning the $244 million, the result is a 0.3% advantage to operating Shoreham" (Robinson, 1988: 68).

Finally, he pointed out that the "no Shoreham" scenarios benefited, in the politically critical short run, from the fact that the costs of the replacement power plants would not "kick in" until the end of the century; "[t]hus, at best, Shoreham's closing might represent not a saving but a postponement of costs; and LILCO's projections, even using Lazard's as the base (without the needed adjustments discussed above), show an economic benefit to running Shoreham over the plant's full 40-year life" (pp. 68-69).

Robinson's opinion, offered to the Public Service Commission, was advisory only. In the second half of 1988 most of the actors involved in the Shoreham controversy, except LILCO, seemed to accept the principle that Shoreham ought not to operate, and were pushing for different alternatives to it. On September 15 the State Public Service Commission approved the settlement in a 6-1 vote. Some commissioners voting for the settlement did so unethusiastically. One, for example, said that "I have struggled to open Shoreham for all these many years. It is impossible. I have to abide by the decision of the people on the Island. It is overwhelming. The possibility of opening Shoreham seems to be nonexistent" (*New York Times*, September 16, 1988: II-2; see also New York State Public Service Commission Opinion 88-29, "Opinion and Order Approving Settlement Agreement," November 18, 1988). Robinson could question the direction of State policy but, given its momentum, not alter it. The next chapter discusses the politics of implementing the settlement, and the alternatives to it.

Appendix to Chapter 6
Chronology of Key Events

1983

June: Consulting firm of Daverman and Associates reports to Suffolk County that acquiring LILCO's assets would cost Suffolk County about $1.4 billion, and would save Suffolk ratepayers between four and thirty percent on electricity bills. Suffolk County Budget Review Office subsequently criticizes report as being based on assumptions "whose validity remains uncertain."

1984

October 23: Ten members of Suffolk County legislature introduce resolution to establish a public electric utility in Suffolk County. The resolution would prohibit the utility from owning or operating a nuclear plant.

1985

September: Hurricane Gloria hits Long Island. LILCO is criticized as parts of Long Island remain without power for eleven days. Chairman William Catacosinos is criticized for not returning from vacation to attend to hurricane crisis.

1986

January: New York State Public Service Commission grants LILCO $68.7 million rate increase, as recommended by its staff. Subsequently LILCO receives extension from banks on payment of loans. Governor Cuomo comments that "If I could start all over with the Public Service Commission, without confirmation, would this be the same Public Service Commission? No."

January: Republican State senators from Long Island endorse concept of a public power authority to replace LILCO on Long Island.

May: Suffolk County Legislature votes to sue LILCO under the Federal Racketeer Influenced and Corrupt Organizations law, charging that LILCO had pursued rate increases from the New York Public Service Commission using arguments it knew to be false.

June: Two panels, one for Suffolk County and one for New York State, evaluate a public takeover of LILCO relatively favorably.

July: New York State Senate (35-26) and Assembly (80-63) agree on bill establishing the Long Island Power Authority (LIPA), with an effective date of January 15, 1987. LIPA is empowered to undertake a friendly or hostile takeover of LILCO if it determines that a public power system would reduce electricity rate increases on Long Island.

1987

January: Governor Cuomo appoints five members to the Long Island Power Authority, with four additional representatives appointed by the leaders of the New York Legislature.

November: New York State Public Service Commission, with five new members and a new Chair appointed by Governor Cuomo in previous year, refuses to grant LILCO a rate increase until Shoreham situation is resolved.

1988

January: Four members of Long Island Power Authority object to the State's negotiations with LILCO--involving other members of LIPA--over a possible settlement of Shoreham in exchange for LILCO's survival as a private utility. The objecting members contend that LIPA's proposed takeover of LILCO is being stalled to accomodate the negotiations.

March 28: The Long Island Power Authority approves an initial "friendly" offer to purchase LILCO's preferred and common stock, and thus assume control over the company.

March 31: Chief State negotiators Vincent Tese and LILCO Chairman William Catacosinos meet in what is later called the pivotal session of the negotiations between the state and LILCO. Tese and Catacosinos agree on the outline of a settlement in which LILCO would abandon Shoreham, and the State would provide enough assistance to allow LILCO to return to financial health.

April: LILCO rejects the Long Island Power Authority's first "friendly" offer to purchase its stock. LIPA raises its bid, but also undertakes steps necessary to initiate a hostile takeover of LILCO.

April: The outlines of a proposed negotiated settlement between LILCO and the State, involving LILCO's selling of Shoreham and financial aid from the State, become publicly known.

May 25: LILCO and the State agree to the settlement.

June and July: The Public Service Commission begins hearings on the proposed settlement, which it must approve. Conflicting studies of the costs of the settlement versus a public takeover of LILCO are presented.

June 22: The Long Island Power Authority votes 6-0 to accept the settlement, with three of Shoreham's opponents choosing not to vote or abstaining.

September 1: Administrative Law Judge Frank Robinson, overseeing the Public Service Commission hearings on the settlement, concludes that the desirable option is to open and operate Shoreham. He also states that, if the Public Service Commission is willing to approve the abandonment of Shoreham, "the settlement is not otherwise objectionable; its rate arrangements are reasonable, and there would be no sound alternative scenarios."

September 15: The New York State Public Service Commission votes 6-1 to approve the settlement.

Chapter 7

THE POLITICS OF SETTLING SHOREHAM

The Reactions to the Settlement

Under the settlement Shoreham would close. By early 1989 seventy percent of the people on Long Island favored this, and 20-25 percent opposed abandoning the plant. Electricity rates would increase about sixty percent over ten years. While some of the increase would occur in any event due to inflation, paying for Shoreham--whether the plant operated or not--was the major factor underlying the rate increases facing Long Island. Governor Cuomo argued that a large rate increase was inevitable, and the settlement allowed the lowest possible increase. But the settlement was rejected both by those who felt that LILCO was "getting off" too easily--the symbol of a "sixty percent rate increase" was powerful--and by those who objected to scrapping a functional 804 MW power plant.

In July 1988 one of four current trustees of the New York Power Authority said he would oppose the settlement because abandoning Shoreham was "not a reasonable energy option" (*New York Times*, June 23, 1988: II-1; *Newsday*, July 5). Long Island's business associations opposed it strongly. In June, the president of the Long Island Association, an organization of businesses, suggested that the costs of providing power to replace Shoreham's were not included in the settlement's projections, understating its true costs, and that "What really sticks in everybody's throat is that you're taking 804 megawatts of energy [from Shoreham] down the tubes. This settlement does everything but provide us with a power supply." The chairman of the 1200 member Advancement of Commerce and Industry objected to the rate increases and questioned whether it would be possible politically to find sites for the five new power plants possible under the settlement.

A *Newsday* poll in early July reported that 71 percent of Long Island residents favored closing Shoreham, 22 percent favored opening it, and 7 percent "[did not] know." However, the respondents split 44-42 on whether they favored or opposed the settlement. More than one-third of those who wanted Shoreham closed still opposed the settlement because it allowed LILCO to continue as Long Island's electric utility, and/or because of the financial benefits granted LILCO. Eighty-four percent of the respondents opposed the rate increases (*Newsday*, July 3, 1988).

Governor Cuomo focussed on the prominent opponents of the settlement relentlessly. He criticized those who favored the plant and opposed the deal as being members of "the extreme right...which has always been for the opening of Shoreham and let safety be damned." He criticized those opposing both the plant and the deal as "the extreme left, which is for a public takeover, even risking the opening of Shoreham." In contrast, he called those favoring the settlement as being "the reasonable middle," and added "To come out against this deal is to come out for the opening of Shoreham, and 71 percent of the people don't want that. I assure you, those who come out against this and then run for re-election are going to have to explain to people why they are for the opening of Shoreham." A Democratic New York State Assembyman from Long Island complained of "heavy-handedness" on Cuomo's part: "When we meet with the governor, or with Richard Kessel, speaking for the governor, what they tell us is: 'You guys have a right to disagree. But, if you do, I have a right to come into your area and correct any misstatements that are being made by you or anybody else'" (*Newsday*, July 5, 1988; July 8).

By mid-July it began to appear that the New York State Legislature might not approve the settlement. Some legislators argued that the agreement required no legislative approval, and that the Governor simply wanted to diffuse responsibility for the future rate hikes. The Governor's aides responded that the New York State Power Authority did not have the authority to build the two gas turbine generators called for by the agreement. Cuomo also suggested that an agreement as important as this one ought to be considered and approved by the State Legislature (*New York Times*, October 3, 1988: II-2).

Cuomo told a press conference on July 14 that Legislative inaction would undermine the agreement and allow LILCO to pursue a license to operate Shoreham at 25 percent of power. He said that "The people of Long Island, overnight, who are confused now about whether they should give a rate increase, will suddenly be confronted by a Shoreham operating at 25 percent...That's one quarter of the way to devastation. And it will be clear that I didn't do that, because I closed the plant with this deal. [The Legislature will] open it by not doing the deal." Cuomo also agreed to modify parts of the settlement. The original preamble to the settlement declared that it was in "the public interest;" Cuomo agreed to drop the preamble, meaning that in approving the settlement the Legislature would not have to wholeheartedly endorse it. A requirement that the Legislature make a "determination of need" for the new power plants--going on record as supporting them--was dropped. A revision of the settlement also specified that LILCO would use competitive bidding procedures in contracting for the new power plants, rather than assuming that LILCO would use competitive procedures. Finally, the original settlement said that the New York State Power Authority (NYPA) would follow its usual procedures when building plants on Long Island, which meant that the Authority could override local zoning laws when siting plants. Cuomo agreed to a change specifying that NYPA could not override zoning laws on Long Island in building the plants.

Nevertheless, on July 16 the Legislature recessed for at least two weeks without having approved the settlement. The Speaker of the New York State Assembly, Mel Miller, had noted that "If there's not a majority of [Long] Island members who are for it and will support it, then we can't do it," and by the recess a clear majority actually opposed it. A *Newsday* count of thirty-one legislators from Long Island indicated that twenty opposed it, four were in favor or leaning that way, and six were undecided or did not state an opinion.

Several factors were behind the legislators' attitudes. Arguably, they did not want to go on record as supporting an arrangement raising electricity rates so sharply. They maintained that the settlement burdened ratepayers heavily and LILCO's stockholders too lightly, and that Cuomo was attempting to spread political blame for an arrangement that was overly generous to LILCO. They also objected to the way the settlement was negotiated without extensive legislative involvement and then presented as what they called a "take it or leave it" proposition. Finally, some legislators argued that a public takeover of LILCO was still the most desirable option (*New York Times*, July 14: II-1; July 16: 29; July 17: 1; July 18: II-1; *Newsday*, July 16; July 17).

On the other hand, outsiders defending the settlement noted that under normal ratemaking the ratepayers should have absorbed the prudent costs of building, and abandoning, Shoreham, and so the rate increases were required legally. Former Public Service Commission Chairman Paul Gioia noted that "State support for the plant, and the County's too, lasted throughout the construction period. And, although a political decision has been made to force the company to abandon the plant, there is still a legal obligation that investors be compensated for any prudent expenditures" (*New York Times*, July 25, 1988: II-3).

LILCO's Progress in Licensing Shoreham

While the settlement was pending LILCO had agreed to not operate Shoreham at more than five percent of power. Still, the company continued to pursue an operating license. This strategy preserved the option to open the plant should the settlement not be implemented. Even if LILCO preferred the settlement, the threat of opening the plant was a valuable resource in its bargaining with the State. By the Fall LILCO's prospects of receiving Shoreham's operating license from the Nuclear Regulatory Commission were good, as LILCO won a series of judicial and administrative decisions moving the plant closer to operation.

On September 6, 1988 a Federal appeals court in Boston upheld the Nuclear Regulatory Commission's revision of emergency planning regulations, issued in 1987. As discussed in Chapter 5, the NRC had stated that it henceforth would presume that state and local governments would assist in any actual nuclear emergency, even if they had refused to participate in emergency planning. Utilities applying for an operating license would have to design and test an emergency response plan; if the NRC judged the plan to be effective *assuming* state and local participation, the plan could be approved. This revision undercut the ability of state and local governments to "veto" operating licenses for nuclear plants through their refusal to participate in emergency planning. The Federal appeals court upholding the revised policy stated that "It is hardly unreasonable for the NRC to predict that state and local governments, notwithstanding their misgivings about the adequacy of a utility plan or their opposition to a particular plant location, would, in the event of an actual emergency at a plant they were lawfully obligated to coexist with, follow the only existing emergency plan" (*Massachusetts vs. United States*, [856 F.2d 378], 383 [1st Circuit 1988]).

A Federal Emergency Management Agency (FEMA) finding in September 1988 also increased the probability of licensing. In June LILCO had run a second test of its emergency plan under FEMA's supervision. In September FEMA's evaluation of the exercise, submitted to the Nuclear Regulatory Commission, reported on the exercise favorably even though state and local officials had not participated in it (United State Federal Emergency Management Agency, 1988; *New York Times*, September 18, 1988: XXI-1).

On September 23, a Federal Atomic Safety Licensing Board, in a 2-1 decision, authorized the Nuclear Regulatory Commission to issue a full-power operating license for Shoreham. In the decision the Board also threw out New York State, Suffolk County, and the town of Southampton as intervenors in the Shoreham case. The Board's opinion noted the history of State and County actions in fighting Shoreham, and concluded that while the actions could be considered "lawful conduct" if considered individually, "in combination they represent a pattern of substantial and continual actions to undermine LILCO's efforts to develop an adequate emergency plan [and] frustrate federal review...Intervenors' actions come perilously close to constituting interference with the federal government's exclusive power to regulate matters of radiological safety...We are forced to conclude that not only are Intervenors unwilling to contribute to the resolution of the important realism issues, but have actively sought to frustrate the Commission's efforts to arrive at an informed judgement" (28 *NRC* 311 [1988], p. 368).

That decision would be appealed, but Cuomo stated that the series of decisions favoring licensing demonstrated the importance of approving the settlement leading to the permanent abandonment of Shoreham. He commented that "When you the clouds darken and you can't see the sun, all you see is dark, heavy clouds, then lightening, and then you get hit with those few drops of rain--those large, heavy drops--it's reasonable to say maybe there's a storm coming." While not endorsing the settlement specifically, Suffolk County legislator Wayne Prospect said that the ASLB's decision meant that "the NRC could try to ram through a license for Shoreham a lot sooner than previously thought," and that "it is absolutely imperative that a practical plan to close Shoreham be adopted now" (*New York Times*, September 25, 1988: 36; *Newsday*, September 25, 1988; September 27).

On October 7, however, an NRC Appeals Board noted that the Atomic Safety Licensing Board issuing the September 23 decision was only one Board considering the Shoreham case--another was still evaluating LILCO's emergency plan--and that no single Board had the authority to dismiss the state and local governments from the proceeding as a whole. The Appeals Board reversed the ASLB's dismissal of the governments and its authorization of a full-power license (28 *NRC* 423 [1988]). LILCO noted that "This is a narrow, procedural issue, which we will appeal to the [full, five-member] commission." A principal impact of the decision, however, was to delay the possibility of full-power licensing until early in 1989 (*Newsday*, October 8).

In October U.S. Secretary of Energy John Herrington wrote to NRC Chairman Lando Zech asking that the Commission "move expeditiously" to license Shoreham, saying that the delay was placing "an enormous burden on our Nation's energy security, the ratepayer, the consumer, and the federal taxpayer." Another Energy Department official commented that "we write letters all the time to commissions around town on issues that affect national energy policy. From that standpoint this is routine and it's an issue of concern to us...We feel (Shoreham's) as important today as it has been over the last few years. I can't give you the impression there's any greater urgency [due to the possibility of the settlement abandoning the plant]. We just need the plant, we need the decision" (*Newsday*, October 12).

Finally, President Ronald Reagan issued an executive order on November 18 which required Federal officials to coordinate emergency response planning when state and local officials refused to participate. The Nuclear Regulatory Commission earlier had specified that it henceforth would presume that state and local governments would respond to emergencies. Reagan's executive order specified that the Federal government *itself* would provide emergency reception and medical centers, communication services, and other facilities if state and local governments failed to do so (53 *Federal Register* 47513 [November 23, 1988]). A spokesman for Governor Cuomo, criticizing the executive order, commented that "Nothing that the Reagan-Bush administration could possibly do to subvert the law and their own policies in the quest to license Shoreham should surprise anybody. That's why we have a settlement. That's why it is now before the legislature. And whether Reagan wins or the people of Long Island win will be determined by the Legislature" (*Newsday*, November 19).

Legislative Action on the Settlement

LILCO's progress in obtaining an operating license softened some of the opposition to the settlement, and it began to appear that the State Legislature ultimately would accept it. On October 26 Suffolk County legislators Gregory Blass and Wayne Prospect, and Nora Bredes--all strong opponents of Shoreham and previously of the settlement--endorsed it as a way to guarantee Shoreham's closing. Blass said that "The settlement by itself strikes a devastating blow to Long Island. However, the operation of Shoreham strikes a fatal blow" (*Newsday*, October 27, 1988).

But the strong anti-settlement campaign carried on by an unlikely mix of Long Island legislators, public power supporters, and Shoreham advocates apparently had affected opinion on Long Island. A *Newsday*/Gallup poll announced on October 30 said that 50 percent of the respondents opposed the settlement, and only 31 percent favored it. Sixty percent of those opposing the settlement opposed it because "It seems that the settlement is too generous to LILCO stockholders at the expense of ratepayers" or "It seems the electric rates will be too high" (Nineteen percent opposed it because "I'd like to see the government take over LILCO" and sixteen percent because "I'd like to see Shoreham in operation"). A similar poll in June reported that 78 percent of Long Islanders trusted Cuomo in the Shoreham settlement negotiations and 20 percent did not; in contrast, the October poll found that only 34 percent approved of Cuomo's performance as a negotiator. The poll also found that at present 62 percent opposed opening Shoreham, 26 percent favored opening the plant, and 12 percent expressed no opinion (*Newsday*, October 30).

The *Newsday*/Gallup poll had noted the rate increase associated with the settlement, but did not suggest that rate increases would differ under different scenarios; New York State had been arguing that rate increases under the settlement would be less than with Shoreham operating. One week later a poll conducted by Governor Cuomo's political committee also reported Long Islanders' attitudes toward the settlement. Unlike the *Newsday*/Gallup survey, the political committee's central question included the statement that "The New York State Public Service Commission has determined that the electric rates if the settlement is approved would be less than if the settlement is not approved and the Shoreham nuclear power plant is allowed to operate." That survey reported that 49 percent of Long Islanders would favor the settlement, and 26 percent would oppose it. The *Newsday*/Gallup poll reported a similar 48 percent "approval" rate when it asked if respondents would favor the settlement if they knew that rate increases would be lower than under the alternatives. A Gallup poll Vice-President commented that "Obviously when you ask a question like that and change the interpretation of one key element, you are going to get different results" (*Newsday*, November 7)

LILCO's shareholders voted overwhelmingly to accept the settlement at their meeting on November 4, 1988. Of the 118 million eligible shares, holders of 82.6 million supported the settlement, holders of 2.3 million opposed it, and holders of the balance did not vote. The New York State Legislature was not scheduled to re-convene until November 28; LILCO's Chairman William Catacosino's said that he was willing to extend the deadline for legislative acceptance of the agreement from November 18 to December 1.

The political and personal effects of the year affected the ostensibly final legislative negotiations. The Governor had not involved the Legislature actively when designing the settlement. When the settlement was criticized as placing too much of the burden on ratepayers, legislators could argue relatively easily that by seeking their approval Cuomo was trying to diffuse responsibility for an unpopular arrangement in which they had not played an important part. Also, targeted legislators and others resented the heavy-handedness of the Governor's lobbying. He had vigorously criticized those who opposed the settlement as either "the extreme right...which has always been for the opening of Shoreham and let safety be damned," or "the extreme left, which is for a public takeover, even risking the opening of Shoreham" (*Newsday*, July 5, 1988). Finally, Cuomo was portrayed as a negotiator who had abandoned "public power" and was enriching LILCO.

For example, settlement opponent Paul Harenberg said that "there's the whole question of tactics and whether it's appropriate for a settlement of this order to be negotiated by the governor's office without input from the Legislature from the word 'go,' and to undercut the legislative intent of the passage of the Long Island Power Authority Act. For that reason alone, it ought to be disapproved, even if it's a good settlement." Another Long Island assemblyman commented that "This inability to say that maybe there are some miscalculations here, maybe this is far too lucrative for LILCO and maybe this is far too burdensome for Long Island's ratepayers--it seems to be absent in his vocabulary." The toughness of the early battle made compromise more difficult than it otherwise might have been (*Newsday*, November 27; December 2).

The uncertainty regarding LILCO's strategy was another important contextual factor in the negotiations. Those backing the settlement maintained argued that LILCO would accept no renegotiation, and that the company intended to open Shoreham should the settlement not be approved. LILCO Chairman Catacosinos told an interviewer that "We negotiated the settlement in good faith, and it's a thin financial transaction to the company--it restores us to health slowly...it will be difficult to meet our obligations to our customers, and the company will not renegotiate the transaction." Vincent Tese commented that Catacosinos "thinks they [the legislators] are out of their tree. He's close to opening the plant, and he's even starting to think that's what they want him to do. He's starting to think they want him to open the plant and get it off their backs." Conversely, the settlement's critics maintained that "the last thing LILCO wants is that plant going on line." LILCO faced serious problems with Shoreham, even with a license, and thus the company arguably could be forced into a renegotiation (*Newsday*, November 27).

In this context of antagonism and uncertainty two issues were particularly important in the legislative discussions. First, legislators disagreed over whether the settlement burdened ratepayers too heavily and stockholders too lightly, and how the settlement could be modified to soften the impact of the deal on ratepayers without driving LILCO out of the agreement. Long Island's legislators argued that the State could funnel relief to Long Island's ratepayers through a special reduction of the State's gross receipts tax (GRT) on LILCO. The gross receipts tax is a tax levied on utilities' revenues; normally, as LILCO's rates rose under the settlement, the taxes collected would rise as well. Cutting that tax for LILCO would diminish its actual costs and thus mitigate the price rise for electricity on Long Island. However, New York then would have to make up the revenue elsewhere, and thus the burden would be shifted to other areas of the state. Cuomo noted that "There can be no gross receipts tax solution that takes away from upstaters. It is possible to do something with GRT, but it will have to benefit Long Islanders without hurting the rest of the state." On November 30 Long Island legislators demanded that $500 million in ratepayer costs be cut from the deal. Senator Ralph Marino, a Long Island legislator who was about to become the Senate majority leader, said that "We think there's a potential for much greater savings for ratepayers...there's the potential for a half-billion or more if the governor and LILCO want to do that for ratepayers" (*Newsday*, November 27; December 1).

Second, what came to be called "the nightmare scenario" was presented as a possibility, raising a new cloud over the deal. As discussed above, in recent months LILCO had won several legal or administrative decisions favoring Shoreham's licensing. Federal law required that the Nuclear Regulatory Commission approve the transfer of Shoreham from LILCO to the Long Island Power Authority, which would then arrange for decommissioning. Legislators opposing the deal argued that, conceivably, the NRC could refuse to approve the transfer, or delay doing so, forcing LILCO to "mothball" the plant and preserving the possibility that the plant could be opened in the future. In that event not only would Shoreham remain, but costs of preserving Shoreham would continue to be paid by ratepayers. It was proposed that the settlement be amended to have LILCO's stockholders pick up the additional costs if the NRC had not acted on the decommissioning plan within three years, but LILCO rejected the amendment (*New York Times*, November 27, 1988: XXI-1; November 28: II-3; November 30: II-6; *Newsday*, November 30).

The Legislature did not act on the settlement by the December 1 deadline. As noted above, the bitterness of the earlier fight over the settlement--a fight conducted during an election--had left compromise difficult to achieve under the best of circumstances. However, the circumstances for decisionmaking were not the best. The few days in late November left to settle complex issues prior to the deadline was too short a period.

The Governor declared that "We're out. The deal's over. Forget about it. Out. Over. No renegotiation. I told you that a thousand times. That wasn't some posture or some tactic. I have said the Legislature was playing Russian roulette. They have now pulled the trigger."

LILCO did not comment publicly on the missed deadline, but the company was expected to continue to pursue the settlement and a license simultaneously. A legislative aide who worked on the negotiations commented that "While there were a lot of issues about the bill and its words, there's a general feeling that LILCO got too good a deal, and it just became synonymous with approving a settlement that a lot of people oppose." Several legislators from upstate New York also said that they could see no reason why the Shoreham plant should not open. Assembly leader Melvin Miller said that "There is nothing ever dead here. I think it's dead for today. That doesn't mean it can't be resurrected tomorrow" (*Newsday*, December 2; *New York Times*, December 2: 1, II-3). Indeed, in a few days an unexpected decision changed the Shoreham situation dramatically.

The RICO Suit

The Initial Verdict and Attempts to Settle the Suit

As noted in Chapter 6, in May 1986 the Suffolk County legislature had voted to sue LILCO, on behalf of itself and all LILCO ratepayers, under civil provisions of the Federal Racketeer Influenced and Corrupt Organizations Law. The County charged that LILCO lied to the State Public Service Commission about Shoreham's progress in order to obtain rate increases. For example, the complaint said that LILCO President Wilfred Uhl testified to the PSC in 1977 that Shoreham would open in 1979, while at the same time writing in an internal company memorandum that it would be "hazardous" to predict that Shoreham could load fuel by December 1979, a step preceding full operation by many months. Had the PSC known LILCO's true progress, the complaint continued, it would not have granted the rate increases it did (*New York Times*, June 1, 1986: 37). During the Shoreham settlement negotiations in early 1988 LILCO insisted that the State pressure the County to drop the suit as a settlement condition. Vincent Tese, the State's principal negotiator, had called the lawsuit "a baloney" issue, and LILCO ultimately dropped the demand.

On September 6, 1988, United States District Court Judge Jack Weinstein, who presided over the lawsuit when it went to trial, told Suffolk County that it could represent only itself in the action and not all of LILCO's ratepayers. In principle, this reduced LILCO's possible liability from about $8.7 billion to $55 million. However, County legislator Wayne Prospect noted that if the County won the case then ratepayers could file for damages and "get back the billions of dollars that we have alleged LILCO has stolen from them." In September Stone and Webster, one of the defendents in the suit, settled with the county for $250,000. Stone and Webster did not acknowledge any of the charges, but said that settling was less expensive than going to trial (*Newsday*, September 7, 1988; September 19; September 24).

The trial before a six-person jury began on October 3. Testimony continued through October and November, with Suffolk introducing what it called evidence of deception and LILCO responding that the Public Service Commission was well aware of the delays at Shoreham. Summary arguments for the case were made on November 29 and 30, 1988. At that time Judge Weinstein indicated that he felt Suffolk's case was weak. LILCO had moved that the case be dismissed, and Weinstein responded, with the jury out of the courtroom, that "I really think this [Suffolk's] case is so thin that it probably can't stand up...[but] I think I am not going to take it away from the jury" (*Newsday*, November 30).

The jury's verdict on December 5 surprised observers and changed the Shoreham situation dramatically. The jury found LILCO and its former president, Wilfred Uhl, guilty of a pattern of "racketeering, lies and fraud" involving the construction schedule, and awarded Suffolk County $22.9 million in damages. Perhaps most important, the decision created the possibility that all 975,000 LILCO ratepayers could be treated as a class that suffered from LILCO's conduct, leaving LILCO liable for over $4 billion in damages and potentially bankrupt.

Judge Weinstein had to decide whether to overturn the verdict and also whether to treat all LILCO ratepayers as a class eligible to receive damages; those decisions were expected in early February 1989. A law professor commenting on the case said that "Verdicts are not lightly overturned, the presumption in the law is that the jurors acted as a reasonable body." Another noted that it was likely that Weinstein would allow the class action suit because the ratepayers constituted a large number of people with a common claim; "[i]t is a standard textbook example for a class action" (*Newsday*, December 7).

Although a verdict was in, both LILCO and Suffolk County had incentives to settle the lawsuit to avoid further litigation. LILCO faced risks that Weinstein would allow the jury's verdict to stand and permit the class action suit. The company also faced risks that an appeal of the jury's decision might fail, and that the NRC might not grant a license to operate a nuclear plant to a utility facing imminent bankruptcy. Thus, LILCO had strong incentives to negotiate some agreement that would assure its survival as an investor-owned utility.

On the other hand, Suffolk County ran the risk that Judge Weinstein might overturn the jury verdict, or that the verdict would be reversed on appeal. Thus, the County had reason to take advantage of its presently strong negotiating position vis-a-vis LILCO. This created the possibility of negotiations among the company, County, and State over a "megasettlement" which would not only limit LILCO's monetary damages associated with the RICO suit but also close Shoreham.

On December 16 Weinstein suggested to attorneys for Suffolk County and LILCO that they try to settle the RICO suit, and offered his assistance. Subsequently Weinstein asked Kenneth Feinberg, a Washington attorney with a reputation as a strong negotiator, to help fashion an agreement settling both the suit and the Shoreham issue in general. Weinstein had a long association with Feinberg, and had worked with him in the settlement between veterans and the firms that had manufactured Agent Orange, a herbicide used during the Vietnam war. Feinberg also knew Mario Cuomo, having chaired a State Commission on sentencing guidelines; Feinberg met with Cuomo in mid-December to discuss the Shoreham situation (*Newsday*, December 17, 1988; December 23).

The Public Service Commission added to the pressure on LILCO to settle its financial problems by refusing, on December 28, to approve a rate increase. The PSC and its staff recognized that the company was in a "grave" financial situation and would likely run out of cash by May 1989 without a rate hike. However, the PSC voted 6-1 to reject any rate increase requests until the key elements of the original settlement were in place--"an end to the Shoreham controversy," a systematic plan for rate increases to return the company to financial health, a plan for both energy conservation and new power supplies on Long Island, implementation of management audit recommendations, and "resolution of all major rate-related items in controversy between the PSC and LILCO." The Public Service Commission noted that "These remain the elements essential to an acceptable settlement from a regulatory standpoint, and we remain willing to approve any settlement that satisfactorily encompasses them. In addition, of course, we must assure that customers do not pay any damages due from the RICO verdict." The PSC then noted that it would "provide a rate increase in the neighborhood of 5% as soon as we have assurance that the other elements listed above are in place. This could mean reaffirmation of the basic settlement agreement at least by LIPA, NYPA [New York Power Authority], and LILCO, who are, with the Public Service Commission, the operative parties to the settlement, or it could mean a revised settlement" (New York State Public Service Commission C88-E-084, December 28, 1988).

One year earlier, in late 1987, the Public Service Commission had denied LILCO a rate increase in order to force a solution to the Shoreham problem; this was one of the factors pressuring LILCO to negotiate the original settlement. Now the PSC was again linking a rate increase to Shoreham's resolution. The denial, added to LILCO's cash flow problems and the potential RICO liability, put enormous pressure on LILCO to settle on any reasonable terms.

Through January 1989, Judge Weinstein, mediator Feinberg, the parties in the RICO suit, an attorney appointed by Weinstein to represent LILCO ratepayers as a class (Judith Vladeck) and State officials continued to negotiate a settlement to the RICO suit. By January 22 Public Service Commission staff members had estimated that LILCO could forego about $400 million in revenue over a ten year period in any settlement and still remain able to secure an investment-grade rating for its securities. As an all day negotiating session on January 22 (Super Bowl Sunday) broke up, Feinberg gave the participants a twenty-point outline of a proposed settlement. The outline contained blank lines instead of dollar figures; however, Feinberg suggested that the parties consider the $400 million figure as a settlement. His proposal allowed Suffolk County to opt out of any settlement with the ratepayers as a class, renewed the original Cuomo-negotiated settlement of Shoreham, setting a deadline for agreement of March 1, 1989, and included a provision deeming the RICO verdict against LILCO "null and void." LILCO's representatives argued that the company could afford only $250 million in a settlement. Suffolk County's representatives declared that they would never accept a renunciation of the RICO verdict (*Newsday*, January 23, 1989; January 24; *New York Times*, January 24: II-2).

Judge Weinstein's Dismissal of the RICO Case

Suffolk County's win in the RICO suit had jolted the Long Island Lighting Company and provided a major impetus for the new negotiations around Shoreham. But on February 11, in a post-verdict ruling, Judge Weinstein threw out the suit against LILCO. He stated that the Racketeer Influenced and Corrupt Organizations Law "does not apply to a rate regulation case such as the one before us," writing that rate setting involved complex technical issues and that the Public Service Commission had "primary jurisdiction" in such matters.

> Prior to the trial the court, based on RICO precedent, was of the view that the federal RICO statute took precedence over state-regulating policy. What the trial proved almost beyond peradventure was that RICO cannot, and should not, be applied in a case such as this to permit a federal jury in a civil case to second guess the rate-making authority of the state...In effect, the jury was asked to retroactively reset the electric rates previously fixed by the P.S.C. with its staff of hundreds of technicians working in such arcane fields as utility rate-making, marketing of utility securities, taxation, economics, generating plant construction, and power grids (County of Suffolk et. al. v. LILCO et. al., Memorandum and Order 87-CV-646 [JBW], February 11, 1989).

Weinstein did not throw out the jury's verdict itself, saying that it was "supportable on [the jury's] view of the facts and the credibility of the witnesses. [While] the evidence of fraud by LILCO and those associated with it seemed to the court to be insubstantial and unconvincing, the jury was entitled to its own view. The court's conclusion is that LILCO's estimates of time to complete Shoreham were made in good faith, but based on misplaced optimism, lack of nuclear experience and events beyond its control" (pp. 9-10). Although Weinstein did not overturn the verdict, by throwing the case out of court he did render the verdict moot unless his action was overturned on appeal.

Weinstein also urged the parties involved to settle both the litigation and Shoreham once and for all, writing that "The court recognizes the difficult emotional, economic, environmental, and political issues surrounding Shoreham. It is not too late to settle these swirling controversies. There is no time better than now for resolution of the entire controversy. Even this case can go on almost indefinitely during protracted appeals which can only benefit the lawyers and put LILCO at further risk. Removal of uncertainty by the parties and officials is essential to the welfare of Long Island." (p. 40).

On the same day as Weinstein's decision throwing the case out of court, Kenneth Feinberg--the Weinstein-appointed mediator--proposed such a settlement, suggesting that LILCO settle the case with ratepayers for $390 million in rate relief over the following ten years, and $10 million in attorneys fees for ratepayers. (Two days later Judge Weinstein certified the class-status of eight ratepayers represented by Judith Vladeck, in effect clearing the way for the class-action settlement). Feinberg also proposed that Suffolk County settle with LILCO by LILCO agreeing to shut down Shoreham and to drop a $500 million suit it had filed against Suffolk charging that the county had overassessed the plant. Under Feinberg's proposal the State would have fifteen days, after the parties signed the settlement, to approve the State's takeover and closing of Shoreham (*Newsday*, February 12; *New York Times*: February 12: 43, XXI-1).

Although Weinstein had thrown the case out of court, LILCO still had incentives to settle to avoid any further litigation, particularly with the ratepayers as a class, for a settlement with ratepayers would diminish the threat of any further appeals. Suffolk County could appeal Weinstein's decision even if the ratepayers settled as a class with the company, but with the ratepayers' claims against LILCO removed the only economic leverage Suffolk would have left was the original $22 million award, not the over $4 billion threat of a class-action award. On February 14 Vladeck approved the $400 million ratepayer settlement, saying that "The appellate chances are limited. In addition, there's the likelihood of litigation for another two years." Seven of the eight class representatives objected to Vladeck's approval; however, Vladeck labeled the settlement "an effort to end this very long, bitter fight." In formally approving the settlement one month later, Weinstein defended Vladeck's action, writing that "class representatives, who may have individual interests that conflict with the interests of the class as a whole, may not block a settlement that is fair and reasonable for the class...The rule is particularly appropriate here, where the objecting named plaintiffs are long-time anti-Shoreham and anti-LILCO activists with interests and political goals that may be antagonistic to the interests of the ratepaying class as a whole" (*County of Suffolk et. al. vs. Long Island Lighting Company et. al.*, Memorandum and Order 87-CV-646 [JBW], 88-CV-2065 [JBW], March 22, 1989, pp. 16-17; subsequently cited as County of Suffolk vs. LILCO, Settlement Approval). Weinstein scheduled three "fairness hearings" for March, during which parties could ask him to approve, modify, or reject the settlement.

The agreement between LILCO and the class of ratepayers isolated Suffolk County in the RICO suit. Mediator Feinberg said that "The train has left the station. Suffolk County can and should be an active participant on any negotiations over the future of the Shoreham facility. If they do not participate in these settlement negotiations, their options are seriously limited on how they can impact on this issue...In light of the agreement, it is inconceivable to me that Suffolk would not make every effort possible to settle this case with LILCO." Nevertheless, that day the Suffolk County legislature voted 18-0 to refuse to abandon the lawsuit unless LILCO promised to "irrevocably close, shutdown, and decommission" Shoreham plant. LILCO immediately called the demand "unreasonable" (*New York Times*, February 15: II-1, II-4; *Newsday*, February 15).

The next day, February 15, the Public Service Commission granted LILCO a 5.4 percent rate increase, but conditioned the increase on LILCO agreeing to not operate Shoreham even if the NRC granted an operating license. LILCO would have to return revenues from the increase if it operated Shoreham and/or if the utility went bankrupt. It was generally agreed that without the increase the company would run out of money by May 1989; thus, the Commission approved the rate increase itself 7-0. The PSC approved the *conditions for the increase*, however, only by a vote of 4-3. Chairman Peter Bradford said that operating the plant at more than the currently authorized 5 percent of power would increase the cost of decommissioning under any settlement, and that the four-member Commission majority wanted to assure that the cost of decommissioning the nuclear plant would not increase if a Shoreham settlement was approved. Also, the restrictions would keep any future operation of Shoreham from "having a disruptive effect on the negotiations" (*Newsday*, February 16, 1989). The day after the Public Service Commission granted the conditional rate increase LILCO's banks assured that the company would avoid bankruptcy in the coming year by approving a $200 million line of credit for the company. The settlement of the RICO suit with LILCO's ratepayers, and the PSC rate increase, made the line of credit substantially safer for the banks (*Newsday*, February 17).

Although a preliminary settlement with ratepayers over the RICO case had been reached, the Suffolk County government still had not settled with LILCO. On February 22 the negotiations between the County and LILCO, mediated by Weinstein and Feinberg, collapsed. Feinberg's latest proposal had revived Governor Cuomo's settlement shutting down Shoreham. The County insisted that the settlement include a strong non-Shoreham clause, but not the Governor's proposal. In addition, the County wanted LILCO to drop its $500 million property tax overassessment suit against the County. LILCO refused both demands. Feinberg suggested that Suffolk County was primarily responsible for the suspension of negotiations, noting that

"LILCO remains willing to accept my language implementing a comprehensive solution to the problem of Shoreham. The Suffolk Legislature is at the present time unwilling, for understandable political reasons, to deal forthrightly with the problems of Shoreham." On February 27 the County Legislature voted unanimously to appeal Weinstein's decision throwing the RICO suit out of court, and also to oppose the settlement with ratepayers. As noted above, undercutting the settlement with ratepayers was critical to the County, for it was the $4 billion threat of a class action award which gave the RICO suit most of its force (*Newsday*, February 23; February 28).

The RICO settlement's fairness hearings were held in March. Attorneys representing the U.S. government, Nassau and Suffolk Counties, Grumman Corporation, the Long Island Railroad, and the Long Island Association suggested that the settlement ought to include funding for an advisory board to monitor LILCO's compliance with the agreement. They also suggested that the settlement's language, by releasing future legal claims arising from the RICO suit, "is so broad it might be used to frustrate legitimate claims against the utility" (*Newsday*, March 9). The Grumman Corporation, Long Island's largest employer maintained that the utility could afford to provide $.75 to $1 billion in rate relief--rather than the $390 million in the settlement--and still avoid bankruptcy.

Judge Weinstein formally approved a final settlement of the RICO suit on March 22 (*County of Suffolk et. al. vs. LILCO et. al.,* Settlement Approval, March 22, 1989). It provided $390 million in rate relief for ratepayers, and $10 million for attorneys' fees. Funding for a thirteen member Citizens Advisory Panel to oversee implementation of the agreement was added (pp. 60-62). Weinstein denied Suffolk County's request for legal fees associated with the case and criticized the County Legislature for not joining in the settlement. Suffolk legislator Wayne Prospect said that Weinstein's "decision is so fundamentally contrary to existing Supreme Court decisions on the [racketeering] statute that the Suffolk County Legislature feels confident we can reverse him on appeal, and that's what we intend to do," and that Weinstein was "trying to intimidate the Suffolk County Legislature into signing the agreement by threatening us with legal fees." County Executive Patrick Halpin, on the other hand, agreed with the judge, noting that Suffolk "has more to lose by not settling and resolving this dispute now than it has to gain" (*Newsday*, March 23). Nevertheless, on November 1, 1989 Suffolk County began its arguments before the Federal Court of Appeals in New York that the RICO case should be reinstated and Weinstein's settlement overturned (*Newsday*, November 2, 1989).

Finally, Judge Weinstein's opinion noted that "In a few years when power and fuel are short and pollution by fossil fuels an even greater problem than it is today, many will rue the day that Shoreham was torn down. But today's decisions must be made on the basis of current fears and calculations, not hindsight. Shoreham is dead. Those who refuse to acknowledge the obvious only make it more difficult to bury the controversy and get on with life. This nuclear plant, in all its technical glory, will be torn down because the people will it" (p. 43).

The settlement negotiated by Weinstein and Feinberg had resolved the RICO suit, barring a successful appeal by Suffolk County. Concurrently the State and LILCO negotiated a revision of the previously failed comprehensive Shoreham settlement.

The Revised Shoreham Settlement

In the first two months of 1989, as the effects of the RICO decision and related negotiations were playing out, LILCO continued to pursue an operating license for Shoreham. At the end of February it appeared that the Nuclear Regulatory Commission imminently would throw New York State and Suffolk County out of the Shoreham proceedings on the grounds that they had worked unfairly to block the NRC's approval of the plant. This would clear the largest remaining obstacle to an operating license. Cuomo noted in an interview that "It looks like we're going to lose" with the NRC, and he and his advisors believed that it would be far more difficult politically to close a fully licensed plant than one awaiting approval.

On February 28, shortly after Judge Weinstein's initial settlement of the RICO suit, Governor Cuomo announced that he and the Long Island Lighting Company had signed a new settlement to close Shoreham. As in the original settlement, LILCO agreed to sell Shoreham to the State for $1, and the Long Island Power Authority would contract with the New York Power Authority for decommissioning. LILCO agreed to not operate Shoreham, even if it received a license to do so, provided that all approvals for the settlement, with the exception of shareholders' approval, were received by April 15. The New York Power Authority would build three new power plants on Long Island, under competitive bidding, if asked to do so by LILCO. LILCO agreed to make a "good faith" effort to settle its tax suit with Brookhaven Town over tax assessments; other litigation over Shoreham also would be ended. The State also would support LILCO's application for $500 million in industrial development bonds (Settlement Agreement--LILCO Issues, February 28, 1989),

The new settlement differed from the original in two major respects. First, the new settlement did not *require* legislative approval, although under a separate arrangement the Legislature would have to modify LILCO's gross receipts taxes to mitigate some of the likely rate increases. Second, the new settlement did not formally guarantee rate increases for

LILCO. The original settlement established a three-year schedule of firm rate increases and seven years of "targeted" rate increases. In contrast, the revision stipulated no guaranteed increases, leaving the size of rate increases solely to the Public Service Commission's discretion. Cuomo indicated that he expected the eventual rate increases to be about the same as in the earlier settlement. However, the new settlement did not specify *formally* any fixed rate increases.

Thus, Cuomo said that the new settlement accomplished the closing of Shoreham "in a way that perfectly satisfies the political needs of the Legislature. They said they didn't want to be accountable. I think that's a mistake by them. I think that is an abdication of their role...Let it for all time be clear that when I closed the Shoreham plant, I did it without the Legislature. And when we build that brass plaque, on the beautiful seeded ground where there would have been a nuclear facility, where children are playing, gently swinging back and forth, let the plaque say no legislators' names." Arguably, the new settlement would be less controversial politically because it did not lock in rate increases. "LILCO wanted a guarantee of what the rates would be, and that we would not and could not give them," Cuomo noted, adding sarcastically that "So, let's leave it all up to the PSC. That gives us all political shelter, and nobody will be responsible" (*Newsday*, March 1, 1989).

Cuomo had backed away from his long-standing demand that the Legislature approve the agreement and announced the settlement individually. The fact that the NRC would probably soon approve Shoreham's operating license helped force a decision. Also, legislators suggested that Cuomo's insistence that the Legislature be involved was beginning to look unseemly. One noted that "I think he felt, 'I'm getting dragged down and I'm looking silly, so I should get above the fray'...I just wish he'd done it a long time ago." Officials involved with the negotiations also pointed out that the Public Service Commission itself had begun considering a deal requiring approval neither by the Legislature nor Cuomo. The PSC, in denying LILCO a rate increase in December 1988, had called for an affirmation of the basic settlement by "the operative parties"--the PSC, LILCO, the New York Power Authority, and the Long Island Power Authority, and "that possibility...raised concern among members of the administration that Mr. Cuomo would be lumped together with the Legislature as too scared to act" (*New York Times*, March 1: 1, II-4; *Newsday*, March 2).

The revised settlement, like the original, received mixed reviews. The leaders of the Republican-controlled Senate and the Democratic-controlled Assembly both supported the new settlement publicly. Those who supported a public takeover of LILCO opposed the new settlement just as they had opposed the old because it removed LILCO's present crisis and left the company as an investor-owned utility.

The Grumman Corporation also called for a mothballing rather than a decommissioning of Shoreham. Its attorney argued that "To bulldoze it to the earth will cost a lot of money. LILCO testified that this summer there are, in all probability, going to be brownouts. There's an energy crisis. It makes no sense, if this plant is going to be licensed, to tear it down when in an emergency situation it might be able to provide energy to the area" The Long Island Association, the largest business association on the Island, commented that "we do not believe the case has been made for closing Shoreham, for abandoning an asset which has already cost the people of Long Island billions of dollars." Seventy-two percent of 340 executives responding to a Long Island Association poll opposed the settlement, and 26 percent favored it. (The same poll reported that the executives opposed the settlement of the RICO suit by a margin of 54-34). Later, the Association said that it planned to intervene before the United States Nuclear Regulatory Commission proceedings to contest the transfer of LILCO's Shoreham license to the Long Island Power Authority, and to ask for a mothballing rather than dismantling of the plant (*Newsday*, March 21; March 22; April 5; May 19).

The revised settlement did not *guarantee* LILCO a package of rate increases. In early March LILCO requested that the Public Service Commission grant it a 4.9 percent rate increase in October 1989, in addition to a permanent granting of the provisional 5.4 percent increase received in January, and rate increases of 4.5 to 5 percent in each of the following nine years. The long-term plan thus called for eleven rate increases, not the ten originally proposed in the Cuomo plan. LILCO also proposed a $1 per common share dividend payment in 1989. LILCO maintained that its property taxes, interest charges, and Shoreham-related delay costs all had increased substantially since the earlier agreement. The company also argued that the proposed rate increases--and the resumption of dividend payments to stockholders--were necessary to return the company to an investment-grade rating, a rating which would eventually lower the utility's borrowing costs. The staff of the Department of Public Service essentially supported LILCO's request.

As expected, Richard Kessel, in his role as Chair of the Consumer Protection Board, opposed the increases, saying that "We could not support an 11th-year rate increase," and proposed a package of ten somewhat lower rate increases, with no additional increases in 1989. Although the original Cuomo settlement--which he had backed as a member of the State negotiating team--included a dividend to stockholders in 1989, Kessel now called for a freeze on common stock dividends for LILCO shareholders until 1991. In late March the Senate Republican leader, from Long Island, opposed granting LILCO the amount requested, saying that "There is widespread expectation on the part of the public that rate increases approved by the PSC will be substantially lower than those established by the governor and endorsed by the PSC last year." Bond rating agencies reacted to this opposition. Standard and Poor's noted that "vocal political opposition [to the rate increases and] consequent uncertainty regarding their ultimate collection" would preclude an investment grade rating "in the near term...Until LILCO can build a financial cushion, or until there is more confidence in its ability to actually implement revenue increases, the current [below investment grade] rating more accurately reflects the risks inherent in its senior secured debt" (*Newsday*, March 11; March 17; March 18; March 23; March 29).

The principal actors in New York State signed off on their parts of the new settlement fairly rapidly. The Public Service Commission approved it on April 5 by a vote of 6-1, and also approved the plan for a ten year series of rate increases of between 4 and 5 percent--in addition to the earlier increase--contingent on the Shoreham settlement being enacted (*Newsday*, April 6).

Under the settlement the Long Island Power Authority would take the title to Shoreham, and the New York Power Authority would arrange decommissioning and build new power plants if asked. On April 12 the New York Power Authority (NYPA), by a vote of 3-2, agreed to decommission the Shoreham plant and build new power plants on Long Island if asked to do so by LILCO. Richard Flynn, the Chair of the Authority and a positive vote, argued that the vote was to carry out a policy decision made by the governor and not an approval or disapproval of Shoreham. However, two of the three positive votes were cast by commissioners appointed by Cuomo during the settlement negotiations and approved by the State Senate in April. Those voting against the agreement maintained that the Power Authority could not take on the tasks the settlement required of it without additional legislative approvals, and also that by participating NYPA risked assuming liability for the Shoreham problem. One commissioner voting against the resolution noted that "LILCO for all intents and purposes is broke and [the Long Island Power Authority] has no assets or net worth...[this leaves NYPA as a] logical target for those seeking redress." Also, he argued in the minority view, the NRC might refuse to transfer Shoreham's license to LIPA and insist that the New York Power Authority assume responsibility for the plant because of its experience in the nuclear field (*Newsday*, April 13). On April 13 the Long Island Power Authority voted 8-1 to support the settlement, with Steven Liss dissenting because of the probable rate increases. The next day the LILCO Board of Directors unanimously approved the settlement, agreeing to recommend approval to the stockholders. On June 28 the company's stockholders approved the settlement, with holders of 82.5 million shares supporting the settlement and holders of 3.6 million opposing it (*New York Times*, June 29, 1989: II-1).

The Federal Government's Opposition to the Settlement

The Federal government has claimed almost exclusive authority to regulate the safety of nuclear plants. This authority, although contested (Campbell, 1988: 81-85), stayed relatively intact until the 1980s. Challenged by the Shoreham case, the Federal government had tried to retain its authority on a number of fronts. The Nuclear Regulatory Commission ruled that it could find emergency plans to be acceptable even in the absence of state and local participation in plan exercises; the NRC and the Federal Emergency Management Agency supported LILCO's exercises of its emergency plan; and President Reagan, through an executive order, directed the executive branch to assist directly in nuclear emergency planning and responses where state and local governments refused to participate.

On March 3, 1989 the Nuclear Regulatory Commission threw out Suffolk County, New York State, and the Town of Southampton as intervenors in the Shoreham proceeding, saying that the governments had "willfully defied the Licensing Board's orders, thereby causing great harm and delay to the efforts of LILCO to demonstrate the sufficiency of its emergency plan...and to the integrity of the Commission's adjudicatory process" (29 *NRC* 211 [1989], 232). The vote was 4-0. (One commissioner had recused himself from all Shoreham votes because of prior legislative work.) On April 17 the Nuclear Regulatory Commission staff reported formally that Shoreham was acceptably safe under NRC regulations, and that LILCO's evacuation plan was adequate, saying that Long Island's geography or other features did not "render emergency planning at Shoreham fundamentally more difficult than at other sites." The director of the Office of Nuclear Reactor Regulation noted that Shoreham's location "compares favorably" with those of other nuclear plants around the nation; that Long Island was "a flat coastal plain without complex terrain," had major roads heading east-west and north-south, and a population density in the area of possible evacuation that was "less than for several other nuclear power plants." Three days later the Nuclear Regulatory Commission voted 4-0 to grant LILCO an operating license for Shoreham. Lando Zech, the Chairman of the NRC, was asked whether he believed it would be a waste "to tear down the plant," and responded that "As chairman of this regulatory commission, that's not my responsibility. As an American citizen, I would call it a waste" (*New York Times*, April 18: II-2; April 21: II-1; *Newsday*, April 18; April 21).

But as the agreement between LILCO and New York State fell into place, numerous actors in the Federal government saw the prospective scrapping of a $5.5 billion, operable facility as a particularly striking and now imminent rejection of nuclear power. Thus, a campaign against the settlement emerged both in the Administration of George Bush and the Congress.

The Administration's Opposition to the Settlement

As the various parties in New York approved the settlement, James Watkins, the United States Secretary of Energy, called it "one of the most foolish deals in the nation's history." Deputy Energy secretary W. Henson Moore said that "We intend to throw up every roadblock we can to it, and if we have to create some, we'll do that." He argued that the plant ought to be operated and, short of that, mothballed rather than decommissioned. The Energy Department and others supporting the plant hoped to delay decommissioning until events such as electricity blackouts, oil embargoes or serious price increases, or election of new state and local officials might favor the opening of the plant. Later, Moore told the Association for a Better Long Island, two business groups opposing decommissioning, that the Administration would attempt to block LILCO's transfer of Shoreham to the Long Island Power Authority (*Newsday*, April 13; April 14; April 16; June 14; June 15; *New York Times*, April 14: II-2).

On August 15 Governor Cuomo met with President Bush and Henson Moore about Shoreham. Moore said afterwards that the meeting changed no minds, commenting that "We found nothing new in the lengthy meeting we had with the governor; there's no reason for us at this point to change our policy and we don't anticipate changing our policy. We offered the governor the option of working with us to mothball the plant, and he refused." Cuomo, in turn, said that it was unclear what position President Bush would take after the meeting; "It was kind of him to allow us time to discuss a problem that is important to three million people on Long Island. It is unreasonable, however, to expect that the president would be able to answer all the questions we had about the position of the Department of Energy....Its obvious he didn't know the facts, and I'm not sure Henson Moore did either, to be candid with you...They made no persuasive argument at all. Frankly, most of the things [Moore] said were inaccurate. I pointed that out and that was the end of the discussion."

Still, on September 1 the United States Justice Department joined a suit in the New York State Supreme Court trying to block the settlement, arguing that the State failed to assess the environmental impact of substituting fossil fuel-fired plants for Shoreham. On November 14 President Bush affirmed what was now a full position of his Administration that it was better to operate Shoreham than to decommission it (*New York Times*, August 16: II-2, XXI-1; Newsday, August 16; September 2; November 15).

The Congressional Reaction to the Settlement

The Congressional reaction to the settlement, outside of the New York delegation, ranged from tolerance to hostility. In April the Senate Committee on Energy and Natural Resources held a hearing on "U.S. Electricity Demand--The Northeastern Region," much of which was devoted to criticizing the settlement. New York Public Service Commission Chairman Peter Bradford defended the settlement, saying that

> Paradoxical as it seems, Shoreham would not have been a great asset to the Long Island power supply situation in the near future. In part, this is due to Long Island's unique circumstances. Because of the transmission constraints, because nuclear units are not reliable in their early years and because Shoreham would be much the largest unit in the LILCO system, much of the generating capacity that is now being built to replace Shoreham would have to be constructed in any case in order to back it up. Furthermore, the higher financing costs associated with utilities with nuclear exposure, to say nothing of the uncertainties caused by continuing to fight about Shoreham, makes the Shoreham-reliance course more expensive than the alternatives.
>
> The fairest summary seems to be that if Shoreham ran well and if oil prices rose rapidly there might come a time when in the next century it would be to the customers net economic advantage. On the other hand, if oil prices remain stable in real terms and Shoreham performs as erratically as many nuclear plants, it would never be an asset economically on Long Island. Given that the near term future without Shoreham is clearly--by LILCO's own estimates--less expensive than the near term future pursuing the Shoreham license, the wiser course is clearly to put the Shoreham struggle behind us and get on with power supply planning on Long Island in a cooperative and constructive manner (U.S. Senate Committee on Energy and Natural Resources, 1989a: 65).

"Senators at the hearing," a report later said, "were incredulous at this claim...that, under the terms of the Settlement, it would be less costly to abandon Shoreham than to operate it" (U.S. Senate Committee on Energy and Natural Resources, 1989b: 1). The following exchange between Chairman J. Bennett Johnston and two utility industry executives, George Edwards and John Eichorn, illustrates the tone:

The Chairman: But you see, they say that they can put [Shoreham] in the rate base, pay for it in the rate base and build new capacity and the ratepayers come out better off than if they went ahead and operated it.

Mr. Edwards: Mr. Bradford is a very smart man, and I am sure he has a good staff. I certainly am not in a position to question his analysis of it until I have read it. I do not know how we could do that...

The Chairman: Is it not really patently absurd to say that that can come about? I mean, honestly, the idea of being able to shut down a plant and put it in the rate base, build new capacity and have it come out less, is that not absolutely absurd?

Mr. Edwards: John, I am going to be around a lot longer than you. I would rather you answer that [Laughter].

Mr. Eichorn: Mr. Chairman, I think you said it quite aptly when you said earlier that the Long Island Lighting Company did not have much of an opportunity to turn down such an arrangement.

The Chairman: I have tried to listen here, and I think maybe I have my ears stopped up or something. Something is wrong somewhere, is it not? What is it? Is there some new principle of hydrogen production in a bottle or something, some new scientific principle that has been discovered that allows that sort of thing to take place?

Mr. Edwards: No, sir, not that I know of. If there is, I am going to buy some of it (U.S. Senate Committee on Energy and Natural Resources, 1989a: 199-200).

In June, Representative Donald Ritter, a Pennsylvania Republican, sponsored an amendment to the Nuclear Regulatory Commission's authorization bill which would prohibit the Commission from transferring Shoreham's license from the Long Island Lighting Company to the Long Island Power Authority. The amendment passed the House Energy and Commerce committee by a vote of 23 to 19. On July 18 House speaker Thomas Foley, deferring to the New York delegation, indefinitely delayed action on the amendment by the full House, but it was debated further in committee and reached the House Floor in November 1989. A full House vote on the NRC's authorization bill, including the amendment, was then delayed until 1990 after Senator Morris Udall (D-Arizona) suggested that the Ritter amendment might consitute a "legislative taking" of LILCO's property (Shoreham), and require the Federal government to compensate LILCO for Shoreham's full $5.5 billion cost. A *Newsday* article commented that "The anti-Shoreham forces on Capitol Hill--composed mostly of New York's lobbyists and its congressional delegation--are very much beholden to Rep. Morris Udall....Those who want to see the nuclear power plant dismantled are fearful that the Ritter amendment might win a fight on the House floor. Therefore, they want a vote on the amendment stalled" (*New York Times*, June 30, 1989: II-1; July 19: II-1; *Newsday*, November 17; November 20; *Watertown Daily Times*, November 10, 1989).

The Economic Evaluation of the Settlement

As discussed in Chapter 6, several economic analyses in New York State had compared the relative costs of (1) a public takeover of LILCO with abandonment of Shoreham, (2) the settlement, and (3) Shoreham's operation. The analyses, which were prepared by those favoring either the settlement or the public takeover, generally shared the view that electricity rates would be lower under either a public takeover or the settlement than under Shoreham's operation. One exception was the opinion of New York State Administrative Law Judge Frank Robinson, who presided over the Public Service Commission hearings on the settlement in June and July. Robinson concluded that it was more economic and desirable to operate Shoreham than to implement the settlement, although the settlement was not otherwise objectionable (Robinson, 1988).

The economic comparison of the settlement versus Shoreham's operation--considered a dead issue in New York--was reopened because of the Federal government's new campaign against the settlement. In July the Senate Committee on Energy and Natural Resources released an evaluation arguing that New York State, through the Public Service Commission, had overstated the costs of operating Shoreham and understated the settlement's costs. The author of the report was its staff economist, Dr. Karl Hausker (U.S. Senate Committee on Energy and Natural Resources, 1989b; New York State Public Service Commission, 1989). This Section will discuss the comparison in detail.

The Public Service Commission estimated that levelized electric rates from 1988-2002, in 1988 cents per kilowatt hour, would be 14.44 cents/kwh under the settlement, and 14.96 cents/kwh with Shoreham in service. Thus, the settlement would lead to an average savings of .52 cts/kwh over the fifteen years (New York Public Service Commission, 1989). Hausker objected to the Public Service Commission's analysis in three areas: that its planning horizon and assumptions regarding financing were questionable; that it did not hold electricity system reliability constant across the comparison, artificially lowering the settlement's costs; and that the PSC's estimates for the cost of capital, fuel, and other factors were biased against Shoreham's operation and in favor of the settlement.

Planning Horizon and Financing

Planning Horizon

The Public Service Commission compared projected electricity rates under the settlement to those under Shoreham's operation for a fifteen year period. The PSC's analysis stated that, until the year 2000, LILCO's rates under the settlement would be lower than with Shoreham in-service. After the year 2000, electricity rates with Shoreham in-service would be lower than under the settlement, "reflecting the substantial drop in rates once the plant is fully phased into rates." However, the Commission said, "By then, ratepayers will have paid about $2.5 billion in rates above the amount they would pay under the settlement. It will take several additional years beyond the turn of the century for ratepayers to break even (New York State Public Service Commission, 1989: 1-2, Chart I).

Hausker responded that it is essential to project relative costs out *at least 20-30 years* when analyzing the costs of power plants with lifetimes of 40 years or more. Capital costs figure heavily early in a project's life, with the investment paying off more substantially in later years. Taking the short time horizon, when the capital costs figured most heavily, would tilt the case against even desirable projects: "If a utility looked ahead only 11 years, it might never build a plant with a relatively high capital cost and lower operating cost (e.g., a nuclear or coal plant)" (U.S. Senate Committee on Energy and Commerce, 1989: 5).

Financing

The Public Service Commission maintained that LILCO could obtain financing more cheaply under the settlement because the company would be a less risky investment; "[B]ecause operation of a nuclear facility is perceived by investors to entail great risks, especially in a hostile environment, the cost of debt and equity is higher."

The Commission said that several features of the settlement would reduce electric rates below what they would be with Shoreham's operation. First, the settlement negotiations "produced a write-off of $250 million of investment beyond the $1.4 billion assessed by the Commission in 1985. The smaller rate base under the settlement reduces LILCO's revenue requirement and rates throughout the study period." Second, under the settlement LILCO's financial risks would drop because the Public Service Commission provided an assured series of rate increases, and the State provided LILCO with at least $100 million per year of low-cost Industrial Development Bonds. With better financial prospects LILCO could substitute debt for equity, dropping its equity share from about forty percent to thirty percent; this, in turn, would allow the company to reduce its after-tax cost of capital because interest on debt is deductible for tax purposes. The settlement would also allow LILCO to refund certain high-coupon debt issues two years earlier than it otherwise would, saving interest payments. In short, the PSC argued, with Shoreham operating LILCO would be a highly risky investment; in contrast, the settlement's financial arrangements provided security which would lower LILCO's financial risks and permit it to finance operations more cheaply. The Public Service Commission estimated that over the fifteen years of the settlement the write-off of $250 million would lower LILCO's rates by about .26 cents/kwh; the lower equity ratio would lower LILCO's rates about .18 cts/kwh; the lower cost of equity and debt would lower electric rates .07 and .03 cts/kwh, respectively; and the early refunding of high-coupon debt would lower rates .04 cts/kwh.

Hausker argued that several of these adjustments were unjustified. The $250 million write-off, which accounted for half of the estimated rate advantage of the settlement over Shoreham, was simply a shift of costs from ratepayers to shareholders; "a sure way to lower rates is to force shareholders to take a greater loss on their investment." The shift had nothing to do, in a technical sense, with Shoreham not operating, could be arranged in a deal allowing Shoreham to operate, and thus was not an "advantage" of the settlement. (Judge Frank Robinson had made a similar point in his recommendation to the PSC in 1988.).

Similarly, the other financial arrangements could be provided as easily, in a technical sense, under Shoreham's operation, and so were not "advantages" of the settlement. The settlement lowered LILCO's financial risks by assuring a package of rate increases, but the PSC could assure rate increases under Shoreham's operation as well. The settlement assumed that LILCO would obtain a minimum of $100 million per year of Industrial Development Bonds for five years, but here again the State could provide such assistance whether or not Shoreham operated. Essentially, the State had made a *political* decision to provide financial aid to LILCO in exchange for LILCO's abandonment of Shoreham. The related financial certainty did not flow from the *economic* merits of not operating Shoreham, but from the political decisions under the settlement to assure rate increases, to shift the costs of providing power from ratepayers to investors, and to issue low-cost bonds (U.S. Senate Committee on Energy and Natural Resources, 1989b: 6-8).

Hausker also argued that the PSC's assessment of the risks of Shoreham compared to the settlement was overstated. While utilities might be charged a risk premium while building a nuclear plant, "there is virtually no risk premium for a utility that simply operates a reactor relative to a non-nuclear utility. The reason is straightforward: during the construction phase, the total cost of the plant is uncertain, as is the amount the regulators will allow into the ratebase. Investors demand a risk premium under these circumstances. Once the plant is complete and ratemaking treatment is finalized, most of this risk disappears" (p. 7). The settlement itself, he added, did not assure LILCO a politically certain future. Given the opposition emerging to the settlement, it was by no means certain that its terms--including a transfer of LILCO's license to LIPA, the decommissioning of Shoreham, and so forth--would fall easily into place.

For these and other reasons, Hausker concluded that "Perhaps one can make a case for up to a .25% risk premium [for debt and equity financing] for LILCO with Shoreham In-Service, as the PSC has assumed. However, there is no basis for making the additional compensations cited above (higher debt leveraging and early bond refunding) for the 'lower risk' of the settlement" (p. 7). Removing from the comparison the $250 million write off of investment (.26 cts/kwh), the lower equity ratio (.18 cts/kwh), and the debt refunding (.04 cts/kwh) narrowed the rate difference between the settlement and Shoreham operation from .52 cts/kwh to .04 cts/kwh. However, this was only the first series of adjustments to the comparison of Shoreham and the settlement that the Senate study recommended.

Reliability

The Public Service Commission acknowledged that, under the settlement, LILCO's electricity generating system would be less reliable than under Shoreham's operation. Thus, Hausker argued, even if the settlement seemed cheaper, the cost comparison was misleading because it compared a less reliable system to a more reliable system.

The New York Power Pool members, including LILCO, were required to maintain electrical system reliability by having a minimum eighteen percent reserve margin over expected peak demand. LILCO itself preferred a higher reserve margin of thirty percent because of Long Island's relative isolation (U.S. Senate Committee on Energy and Natural Resources, 1989b: 9). If Shoreham operated at 65 percent capacity, as the PSC's study assumed for analytical purposes, LILCO would exceed these reserve requirements for the foreseeable future. In contrast, power supplies under the settlement were problematic; in five of seven years from 1989 through 1995 LILCO would violate even the minimum eighteen percent reserve requirement. The Public Service Commission itself noted that

> The energy supply plan for the settlement does not assume the construction of sufficient capacity on Long Island to meet the near-term projected requirements. Between 550 and 1150 mW of capacity are assumed to be available for import from the New York Power Pool.

Such an assumption may be feasible, but raises several questions which apparently were not addressed. 1150 mW approaches the maximum capacity of LILCO's share of the ties to Long Island. None of the upstate utilities has plans to construct capacity to sell to LILCO, so there is no assurance that the capacity will be available for sale. If upstate loads continue to grow strongly, the upstate utilities may themselves need capacity earlier than past projections indicate. This could result in earlier construction of supply options on Long Island or the allocation of more Canadian power. Moreover, the import of 1150 mW will present a difficult operating problem caused by the potential loss of transmission. It is also not clear in reviewing the technical tables that adequate funding was included for the capacity assumed to be available from upstate. The numbers, however, were prepared by LILCO.

The Shoreham-in scenario does include enough capacity on Long Island to supply the projected load without the reliance on the tie to upstate. Shoreham itself raises a problem because of the size of the unit compared with the system. Nevertheless, the reliability of the Shoreham-in scenario should be greater than the settlement scenario. The ties to other systems will be available for emergency support (e.g., the loss of Shoreham) rather than needed to support load under normal conditions (New York Public Service Commission, 1989: 8-9).

The reliability problems would be worsened if demand for electricity grew faster than predicted, and the predictions of load growth made by the settlement in fact were much lower than the growth experienced in the previous few years. The Public Service Commission, based on its review of LILCO's projections, forecast that baseline average annual load growth would be 1.8 percent from 1988-2002, and that this would be reduced to 1.64 percent by "aggressive" demand side management (or conservation). Yet, LILCO's average peak load growth from 1980 to 1988 was 2.5 percent, and over 3 percent per year after adjusting for weather; and in 1987-88 weather-normalized demand growth was 5-6 percent. This pattern of load growth was consistent with that elsewhere in the Northeastern U.S. over the period. The PSC, which acknowledged that load growth might be higher than predicted, maintained that independent power producers could produce more than expected and possibly "offset errors in the estimates of load growth and conservation." But the judgement appeared to be a soft one based only on how the agency assessed early interests in the independent power production program (pp. 4, 7-8; U.S. Senate Committee on Energy and Natural Resources, 1989b: 11).

The Public Service Commission argued that Shoreham could cause reliability problems as well. Shoreham would be an 809 megawatt share of LILCO's 4600 megawatt system. There was the possibility that the plant would perform as poorly as New York's Nine Mile Point II plant, which operated less than one third of its first year. Thus, as PSC Chairman Peter Bradford suggested, much of the additional electrical capacity under the settlement might be needed in any event to back up Shoreham (U.S. Senate Committee on Energy and Natural Resources, 1989a: 65; New York Public Service Commission, 1989: 2).

The Nine Mile Point II plant, however, was exceptional. Thirty-one nuclear plants had begun operation since 1983, and their average capacity factor through 1988 was 68 percent; the average *first year* capacity factor for the plants was 64 percent. Industrywide, capacity factors for all nuclear reactors in 1988 was 63.5 percent. Hausker argued that "the risk of Shoreham being a poor performer is real but small. It is not fundamentally different from the other risks facing LILCO regarding load growth, the success of [conservation], fuel availability, independent power producers, etc." (U.S. Senate Committee on Energy and Natural Resources, 1989b: 12; *Monthly Energy Review*, September, 1989: 89). As noted above, the PSC had acknowledged that "...the reliability of the Shoreham-in scenario should be greater than the settlement scenario. The ties to other systems will be available for emergency support (e.g., the loss of Shoreham) rather than needed to support load under normal conditions" (New York Public Service Commission, 1989: 9).

Adding to the settlement's costs enough generating capacity to bring LILCO up to the eighteen percent reserve margin required by the New York Power Pool would increase average electric rates under the settlement by .04 cts/kwh. Hausker included these additional costs in a revised comparison of Shoreham in-service to the settlement, discussed below.

Hausker also suggested that the settlement continued LILCO's total dependence on oil and gas for power generation; "Diversity in fuels lowers the risk that any one event can have a dramatic effect on a utility's costs...LILCO is extremely vulnerable to a sharp escalation in oil and gas prices and an oil supply disruption. The Settlement would not change this. If Shoreham operates, LILCO's oil and gas dependence would drop to 83%. The value of this increase in fuel diversity is hard to quantify, and no attempt to do so is made here. Nonetheless, the Settlement carries the hidden cost of maintaining LILCO's complete lack of fuel diversity" (U.S. Senate Committee on Energy and Natural Resources, 1989b: 14).

Maintenance, Oil, and Other Costs

New York's comparison estimated certain costs for both Shoreham's operation and the settlement. The Senate study maintained that several of these estimates were "unreasonable... In every case, the unreasonable estimates used by the PSC help make the settlement appear to be the cheaper option" (p. 14).

Operating and Maintenance Cost

The Public Service Commission estimated that Shoreham's operating and maintenance (O&M) cost would be $129 per kilowatt in its first full year, increasing 1.8 percent for most of the study years (Hausker adjusted all figure to 1989 dollars; p. 15). This was above other estimates. In 1987, O&M costs averaged $85 per kilowatt for a reactor of Shoreham's size. The Department of Energy estimated future O&M costs to be about $100 per kilowatt for a reactor of LILCO's size. A model of costs by the National Economic Research Associates (NERA) estimated Shoreham's O&M costs to be $114 per kilowatt.

The PSC estimated that *capital additions* to Shoreham in its first full year would be $93 per kilowatt to be added to the rate base, escalating 1.8 percent for most of the study years. Again, the three sources of information projected that capital additions would cost less than the PSC estimated. In 1987 the industry average for capital additions was $23 per kilowatt; the Department of Energy estimated future costs of $64 per kilowatt, and NERA estimated $63 per kilowatt. Furthermore, for the period 1981-1987, boiling water reactors had capital additions averaging $35 per kilowatt per year; Mark I reactors averaged $47 per kilowatt per year, and Mark II reactors (Shoreham's type) averaged $20 per kilowatt per year. Hausker suggested that

> The [electricity] rate for Shoreham In-Service decreases if one uses more realistic assumptions concerning O&M costs and capital additions. The estimates cited above based on the NERA models and from DOE are conservative in that they are significantly above industry averages. When the NERA or DOE estimates replace the PSC assumptions, the levelized [electricity price] rate for Shoreham In-Service decreases by .12 and .15 cts/kWh, respectively. The leveling off of these costs in recent years is reason for a guarded optimism about the industry's ability to keep these costs manageable in the future, hence this study adopts the higher DOE figure. Shoreham In-Service is the economic alternative using either figure (U.S. Senate Committee on Energy and Commerce Committee, 1989b: 17).

Future Price of Oil

The future price of oil was a critical variable in the comparison of Shoreham and the settlement. Higher oil price increases would increase Shoreham's economic value because Shoreham substituted nuclear fuel for oil; conversely, lower oil price increases would diminish Shoreham's value. The Public Service Commission forecast that oil price increases would not exceed a general inflation rate of four percent through 1991, and would rise two percentage points above a four percent inflation thereafter; in other words, that oil prices would rise 0 percent in real terms until 1991, and two percent thereafter. However, less than one year earlier the Public Service Commission had forecast that oil prices would rise, *in real terms*, 1.5 percent through 1994, four percent in 1995, and 6.5 thereafter, in sharp contrast to its current estimates of zero percent real price rise through 1991 and 2 percent thereafter (U.S. Senate Committee on Energy and Commerce, 1989b: 17). The PSC commented that its current "forecast differs somewhat from the estimates used by the Commission in its recent long run avoided cost opinion...The continuing softness in the oil markets supports a lower forecast in the early years and possibly lower increases through the 1990's. The uncertainty about oil prices in the mid to late 1990's is large; tightening markets could result in higher than forecast increases. Given a real price increase of 2% beginning in 1992 the forecast does not seem unreasonable" (New York Public Service Commission, 1989: 5).

The Senate study suggested that the forecast was quite unreasonable. The new PSC forecast was substantially lower than the forecast of three other reputable sources: the Department of Energy, Data Resources Incorporated (DRI), and the WEFA Group. For example, for 1998 the Public Service Commission forecast that oil would cost $2.50 per MMBTU (thousand British Thermal Units). In contrast, WEFA forecast a price of $3.20, Data Resources $3.60, and the Department of Energy about $4.10 (U.S. Senate Committee on Energy and Natural Resources, 1989b: 18). Hausker argued that it would be more reasonable to substitute the middle-range of the estimates--which was DRI's--for the PSC's. The effect was that the higher oil price "would increase rates for both the Settlement and Shoreham In-Service, but the rate increase is higher in the Settlement due to the larger volume of oil burned." The increase in levelized rates was 0.54 cts/kwh and 0.40 cts/kwh for the settlement and Shoreham in-service, respectively (p. 19).

Cost of Settlement's Conservation Programs

The settlement, according to the PSC, "presumes that LILCO will make an extraordinary effort to reduce consumption in its service area." If it would cost far more than expected to make the necessary reductions in demand, the settlement would be less attractive economically. In fact, the estimated cost of the conservation programs seemed to be excessively optimistic.

Both the settlement and Shoreham-in scenarios included conservation programs costing about $11 million in a first year (increasing thereafter at the inflation rate) and reducing consumption by about 20 megawatts in each year. The settlement added conservation expenditures of an additional $30 million per year, and this "aggressive" program was expected to reduce demand by about 280 megawatts of capacity. From 1990 to 2002, the average cost of reducing demand for generating capacity by one kilowatt under the *base* plan was $761; the average cost of saving a kilowatt of electricity under the *aggressive* plan was $146 (New York State Public Service Commission, 1989: 5; U.S. Senate Committee on Energy and Natural Resources, 1989b: 19).

The "puzzling" aspect of this, Hausker argued, was that the "aggressive program appears to get more bang for the buck," when the principle of diminishing marginal returns would suggest the opposite. One would expect that LILCO's base plan would target the easiest means to conservation, and the aggressive plan, next, to target the somewhat harder problems of conservation. Accordingly, one would expect the first tier of benefits to cost less on average than the second tier of benefits on average. However, the opposite was the case here; the "base" plan reduced demand at an average cost of $761 per kilowatt, while the "aggressive" plan reduced demand at an average cost of $146 per kilowatt. Thus, the cost for the settlement's conservation program seemed to be understated, raising the problem that either the settlement's cost would increase by some unknown but possibly significant amount, or that the conservation targets would not be met, further lowering electrical system reliability under the settlement.

Book Life of Investments and Decommissioning

A "minor but troubling flaw" was that the Public Service Commission treated the "book life" (the number of years over which an investment is amortized) of two investments much differently under the settlement than under Shoreham's operation, making the settlement appear cheaper.

The PSC's analysis focussed on LILCO's projected revenue requirements for the period 1988-2002. *For the settlement*, the PSC spread the costs of a gas turbine and a gas refiring project over forty years, lowering LILCO's projected revenue requirement for the period. In contrast, for the "Shoreham In-Service" scenario the PSC set the book life of the same investments at *fifteen years*, concentrating their costs in the study period and making Shoreham seem to be the more expensive option. Setting the book life of the investments at twenty-five years would increase the settlement's costs by .01 cts/kwh and decrease Shoreham's costs by .01 cts/kwh. Hausker commented that "One hopes that any remaining, undetected errors in the PSC analysis are not significant and are neutral in their effect on the rate comparison" (p. 20).

Finally, as the PSC had noted, its study had not included the costs of a decommissioning fund for Shoreham, a fund now required by the NRC. Including such a fund to the Shoreham In-Service scenario added .01 cts/kwh to its costs (U.S. Senate Committee on Energy and Natural Resources, 1989b: 20).

The Revised Comparison of the Settlement and Shoreham

The Senate study presented a revised comparison which corrected what it called "errors and biases" in the Public Service Commission's analysis; its conclusion was that electricity rates under Shoreham's operation would be lower than under the settlement. The adjustments in Table 16 represent increases in the PSC's estimates; the parenthesized figures represent decreases.

Another way to compare the two policy options was to compare the *real incremental economic costs* of the settlement versus Shoreham. In the *overall electric rate* comparison, the incremental costs were lumped in with all of LILCO's other costs, obscuring their magnitude. In addition, tax treatment and accounting conventions affected the comparison of electric rates under the two scenarios. In contrast, "real economic costs (or economic opportunity costs)...are measured by the actual outlays the utility makes for capital items, fuel, O&M, etc...They are unaffected by accounting conventions or tax rates" (U.S. Senate Committee on Energy and Natural Resources, 1989b: 22).

Under the settlement LILCO would pay for Shoreham's decommissioning, meet near-term power needs with increased oil-fired generation and purchased power, build combined cycle generating plants in the late 1990s, pursue aggressive conservation programs, and undertake several other projects. With Shoreham in-service, LILCO would pay for the O&M costs and capital additions for the plant, pay for nuclear fuel costs, and build gas turbines in the late 1990s.

The study projected the real economic costs of the settlement versus Shoreham in two ways--using the Public Service Commission's original assumptions, and also using the revised assumptions discussed above. As noted, Hausker substituted Data Resources Incorporated's oil price forecast for the PSC's, added the costs of a gas turbine plant for increased reliability under the settlement, and used the Department of Energy's estimates for O&M and capital additions at Shoreham. In addition, in the projection he reduced the costs for the construction of combined cycle plants in the settlement, and dropped from the Shoreham scenario the costs of a coal plant due on line in 2010 (noting that "There is no discernible reason why it is in the Shoreham In-service but not in the Settlement") (p. 23). All remaining LILCO costs were common in both plans.

Using the Public Service Commission's assumptions, the real economic cost of operating Shoreham was about $50 million less than the settlement, a difference of about three percent. Using the revised assumptions, the real economic cost of operating Shoreham was about $750 million less than the

Table 16. Revised Estimates of Levelized Electric Rates, 1988-2002 (1988 cts/kWh)

	Settlement	Shoreham In-Service
ORIGINAL PSC ESTIMATES	**14.44**	**14.96**
Adjustments re capital structure		
Exclude $250 million writeoff	.26	
Exclude lower equity ratio	.18	
Exclude delay in refunding high coupon debt	.04	
Additional gas turbines for settlement's reliability	.04	
Lower nuclear O&M and capital additions		(.15)
Higher oil prices	.54	.40
Correction in book life of of investments	.01	(.01)
Decommissioning account		.01
REVISED ESTIMATES	**15.51**	**15.22**
SAVINGS FROM SHOREHAM IN-SERVICE		**.29**

Note: Totals do not add due to rounding

SOURCE: U.S. Senate Committee on Energy and Natural Resources, 1989b: 21.

settlement, more than a third of the cost of the alternative (see Table 17). This estimate, arguably, was conservative:

These revised estimates...give the benefit of the doubt to the settlement. The time horizon of the analysis is still relatively short (15 years) when a period of 20-30 years would be more appropriate. Reliability is still lower in the settlement....No value is assigned to the improvement in fuel diversity for Shoreham In-Service. An increase in oil prices is assumed to have no effect on purchased power costs when, in fact, there would be some effect. Finally, the PSC assumptions on the cost of [demand side management] programs are accepted as is, although it seems likely that they underestimate the true cost...

In sum, power from Shoreham will not come cheaply, but it will still be less costly than oil-fired generation, power imports, and the construction of replacement capacity. The New York PSC's cost comparison is flawed. An objective assessment of costs shows Shoreham In-Service to be the more economic alternative (U.S. Senate Committee on Energy and Natural Resources, 1989b: 24).

The Denouement at Shoreham

The Federal government's opposition to the settlement did affect the late politics of Shoreham, but at that point the obstacles to opening the plant, or even to mothballing it for possible use in the future, were overpowering. The major problem facing Shoreham's proponents was that the Long Island Lighting Company had signed an agreement assuring the company's financial survival in exchange for its abandonment of the facility. Because LILCO lacked a commitment to open the plant--in fact, had signed a commitment to not open the plant--it would not invest the resources necessary to comply with new regulations by the NRC, or invest anything beyond the minimum maintenance required by its current license. Chairman Peter Bradford of the PSC noted that "The Nuclear Regulatory Commission can't compel LILCO to spend money on a plant it has no intention of operating. The whole regulatory apparatus is license suspension and license revocation. And in this case, it would make LILCO's day."

In mid-July 1989 LILCO began removing the nuclear fuel from the reactor. On September 19 LILCO asked the NRC to formally bar the company from operating Shoreham, allowing LILCO to further reduce its maintenance costs for the plant. In December LILCO requested permission to disband its emergency planning force, in view of the fact that the company did not intend to operate the plant. Richard Kessel, who was now Chairman of the Long Island Power Authority, said in September that "One month ago LILCO and LIPA were not really even talking to each

Table 17. Present Value of Real Economic Costs (Millions of 1988 Dollars)

	Using PSC Assumptions	Revised Estimate
SETTLEMENT		
Shoreham Decommissioning	$381.4	$381.4
Fossil Fuel and Purchased Power	735.3	924.3
"High Production" O&M Costs	23.7	23.7
Combined Cycle Plants--Construction	494.4	387.6
Combined Cycle Plants--O&M	48.8	48.8
Gas Turbines--Construction	n.a.	74.0
Aggressive Demand Side Management	242.4	242.4
Misc. Capital Spending	96.9	96.9
Total	**$2022.7**	**$2179.1**
SHOREHAM IN-SERVICE		
Shoreham O&M and Capital Additions	$1687.9	$1158.1
Gas Turbines--Construction	89.4	89.4
Gas Turbines--O&M	6.4	6.4
Coal Plant--Construction	11.6	n.a.
Nuclear Fuel	177.7	177.7
Total	**$1972.9**	**$1431.6**
SAVINGS FROM SHOREHAM OPERATION	**$49.8**	**$747.5**

Note: Totals may not add due to rounding

Source: U.S. Senate Committee on Energy and Natural Resources, 1989b: 23.

other. Since then we've opened up communications...and I believe that LILCO is being responsive. I am satisfied, at this particular point, that the Long Island Lighting Company is taking actions appropriate with the agreement while at the same time meeting all the obligations of its operating license."

The Nuclear Regulatory Commission said that, while it would require LILCO to adhere to the terms of its operating license (and not let the plant become a "rust bucket"), it would not erect any unusual regulatory barriers to the settlement. The NRC also told LILCO, however, that there was as yet no "clear path" to Shoreham's decommissioning, casting some doubt on whether the NRC would allow LILCO to transfer its license to the Long Island Power Authority until the decommissioning plans were set. The decommissioning of such a large facility had never been done in the United States. The fear that the NRC would delay the transfer of the license to the Long Island Power Authority had been raised in the New York Legislature the previous year; the "nightmare scenario" was of LILCO being forced to retain Shoreham indefinitely, yet not being able to operate it (*New York Times*, June 30, 1989: II-1; October 12: II-7; *Newsday*, July 29; September 20; September 28; September 29).

By early November an option which might break the remaining obstacles to the settlement surfaced--converting Shoreham to a gas-fired plant. It was agreed generally that Shoreham could be converted, but also agreed that a converted plant would provide power far less efficiently than building new plants. The steam produced by a gas-fired plant, for example, had a much higher temperature and pressure than steam produced by a nuclear plant, requiring either a lowering of steam pressure and temperature (reducing the efficiency of the converted plant) or a major modification of the plant's turbine. One engineering estimate indicated that "a converted plant would use twenty-seven percent more fuel than a new combined-cycle plant, potentially offsetting any savings achieved through the use of the plant's existing equipment" (*Newsday*, December 3).

But, efficiency considerations aside, converting Shoreham was attractive *politically*. Any community selected as the site for a new, large power plant would probably try to block it. In contrast, the Shoreham facility already existed, and the Shoreham-Wading River School District depended heavily on the tax revenues drawn from Shoreham (In fact, the District opposed the settlement abandoning the plant). Thus, Shoreham-Wading River would likely accept a converted plant. Furthermore, the Federal government, local business groups, and others trying to block the settlement might ease their opposition if the Shoreham plant was to be used, somehow, rather than dismantled. Using the facility would soften the politically powerful image of tearing down a new, licensed nuclear power plant (*Newsday*, December 3, 1989). On January 3, 1990, in his State of the

State speech, Governor Cuomo said that "Under the leadership of its chairman, Richard Kessel, LIPA will push forward aggressively in its efforts to decommission Shoreham. LIPA will be assisted in this effort by the [New York] Power Authority. Meanwhile, I have asked LIPA and the Power Authority to examine thoroughly the technical and economic feasibility of converting Shoreham to gas" (Governor's State of the State Message, 1990, p. 137).

Conclusion

The argument for the settlement was that it was a good first step in moving away from the Shoreham problem to a situation in which Long Island had a reliable, if unavoidably expensive, supply of electricity. LILCO would abandon Shoreham, and thus the nuclear power plant would be gone from Long Island. With that most controversial issue removed, the efforts to plan Long Island's energy future presumably would be somewhat less contentious and more manageable.

But the company emerged from the settlement about as well off financially as it reasonably could expect. LILCO would recover most of the prudently incurred costs of the plant, just as it would have if the plant had operated. With the Public Service Commission's approval, LILCO would receive a long-term package of rate increases of between 4-5 percent in each of the next ten years. LILCO would need to pay for the decommissioning of Shoreham, but it would recover those costs through higher rates. The State would assist LILCO in meeting the demands for power on Long Island--the costs again recovered through rates--and also in returning LILCO's securities to an investment grade rating. The State justified the financial help on the grounds that an investment grade rating would lower LILCO's borrowing costs in the future, and in the long run reduce electricity rates below what they otherwise would be.

Shoreham's operation did not assure LILCO's survival as a financially stable investor-owned utility because of the extremely hostile political situation in New York State and Suffolk County. In contrast, the agreement came close to providing such a guarantee. Although the company "sacrificed" Shoreham, arguably it emerged better off than it would have with Shoreham's operation in the current political environment. Thus, the State and local officials and the groups on Long Island opposing Shoreham succeeded in blocking the nuclear facility, and LILCO negotiated its way out of potential bankrupty.

Yet, the case does not seem to be a success story. Politics almost always plays an important part in technical and economic decisionmaking. However, in this instance the disparity between the technical and economic evaluations of the plant and the policy of the State was so large as to be troubling. The technical and economic evaluations--with the exception of New York State's evaluations--generally supported the idea of operating the plant, given that Shoreham was already built. The arguments that emergency planning for Shoreham was impossible, that the plant was "unsafe," or that LILCO was incapable of operating it when it was ready for licensing were reviewed in unprecedented length and detail in courts, Federal and State proceedings, and ultimately the Nuclear Regulatory Commission. Fairly consistently, and across a range of judicial, administrative, and technical bodies, the decisions were that emergency planning for Shoreham was feasible, and that what critical problems existed were due primarily to State and local refusal to participate in the planning process; that the plant was acceptably safe by any reasonable standard; and that, after testing, LILCO had shown itself capable of managing Shoreham.

Suffolk County and New York State focussed on the emergency planning issue in opposing an operating license. In dismissing the County and State from the case before the Nuclear Regulatory Commission (a dismissal reversed by an Appeals Board on procedural grounds, but affirmed by the Nuclear Regulatory Commission itself), an Atomic Safety Licensing Board said in 1988 that

> Recent testimony from Dr. Axelrod, the State's top official responsibile to the Governor on disaster and emergency matters, suggests that the Governor's decision to oppose the Shoreham facility was not based on a State technical evaluation of an emergency plan for Suffolk County. The Applicant [LILCO] is entitled to explore, therefore, through State witnesses, whether the decision may have been more of a political edict as the testimony now standing implies.
>
> ...Throughout the protracted period of this proceeding, Intervenors [Suffolk County and New York State] have provided little evidence of a motivation to have this controversy (whether or not adequate emergency plans meeting NRC regulatory standards could, or could not, be developed for the Shoreham facility) resolved on the merits, and in a timely manner. They have chosen to ignore the licensing Board's decision in 1983 which stated "If the County seeks to have its findings adopted, [the inadequacy of LILCO's emergency plan and the nonfeasibility of developing adequate emergency planning for Shoreham], it must litigate before us the facts which it believes support its view that it is not feasible to implement emergency preparedness actions which would meet NRC regulatory requirements in the event of a radiological emergency at the Shoreham nuclear power plant" (17 NRC 608, 643 [1983]). In the place of presenting a positive case to evidence the nonviability of an emergency plan, Intervenors instead have persistently relied on statements of noncooperation with the Applicant and on County resolutions and policy statements that an adequate emergency plan was not possible. This persistency has been in the face of NRC statements and federal case law that the adequacy of emergency planning is the jurisdictional responsibility of the Commission (28 NRC 311, 365-367 [1988]).

The Administrative Law Judge overseeing the settlement hearings for the New York State Public Service Commission concluded in 1988 that the settlement "is not in the public interest because it would scrap a completed $5.3 billion power plant whose operation should save Long Island money, would avoid a real danger of power supply deficiencies, and would be safe, clean, and environmentally beneficial; and that the settlement's benefits are not a sufficient tradeoff, because the power supply arrangements would be moot if Shoreham operates, and its rate arrangements can be (and would have to be) implemented independently" (Robinson, 1988: 4. In a footnote he commented that "Stronger language has been employed: The *New York Times* has called this settlement 'preposterous' [May 13, 1988] and the U.S. Secretary of Energy has labelled it a 'travesty,' both for reasons similar to those put forth herein" [p. 48]).

Federal Judge Jack Weinstein, who heard Suffolk County's RICO suit of LILCO and approved the settlement of the RICO suit between LILCO and the ratepayers, wrote that "The evidence demonstrated that, had Suffolk and the state cooperated with LILCO, Shoreham would now be in operation and LILCO and its shareholders and ratepayers would be in a more favorable economic position. The cost to LILCO and others of Suffolk's and the state's change of views regarding Shoreham cannot be precisely measured. It can reasonably be estimated as at least in the hundreds of millions of dollars" (*Suffolk County v. Long Island Lighting Company*, U.S. District Court [Eastern District], memorandum and order 87-CV-646, p. 6). Essentially, while New York State and Suffolk County won the arguments over Shoreham's safety and economic value in their own inner circles, they lost the arguments virtually everywhere else.

Beginning in 1982, State and local policy toward Shoreham was based on the simple core belief, and the politics of the belief, that the plant was unacceptably dangerous. Basing public policy on this belief cost society about $750 million more for electric power over the next fifteen years than it otherwise would have to pay. Although virtually every reasonably neutral reviewing body rejected the State and local position, what the governments had in their favor was the ability to wear LILCO down financially and politically. The next Chapter makes a final assessment and evaluation of Shoreham's history, and actors' roles in it.

Appendix to Chapter 7
Chronology of Key Events

1986

May: Suffolk County Legislature votes to sue LILCO under the Federal Racketeer Influenced and Corrupt Organizations law, charging that LILCO had pursued rate increases from the New York Public Service Commission using arguments it knew to be false.

1988

April: The outlines of a proposed negotiated settlement between LILCO and the State, involving LILCO's selling of Shoreham and financial aid from the State, become publicly known.

May 25: LILCO and the Governor's Office agree to the settlement.

July: A *Newsday* poll reports that the public on Long Island divides 44-42 percent on whether they favor or oppose the settlement. The majority of those opposing the settlement do so because of the rate increases the settlement guarantees to LILCO.

July 16: The New York State legislature recesses for at least two weeks without having approved the settlement. Strong political opposition to the settlement by Long Island legislators begins to emerge.

September 1: Administrative Law Judge Frank Robinson, overseeing the Public Service Commission hearings on the settlement, concludes that the desirable option is to open and operate Shoreham. He also states that, if the Public Service Commission is willing to abandon Shoreham, "the settlement is not otherwise objectionable; its rate arrangements are reasonable, and there would be no sound alternative scenarios."

September 2: The Federal Emergency Management Agency concludes that LILCO's emergency response plan would provide reasonable assurance of public safety in the event of an accident at Shoreham, even though state and local governments did not participate in the plan's exercise.

September 6: A Federal Appeals Court in Boston upholds the Nuclear Regulatory Commission's revision of its emergency planning regulations, allowing the NRC to presume that state and local governments will participate in an actual emergency response even though they have not participated in plan exercises.

September 6: Judge Jack Weinstein tells Suffolk County that it can represent only itself, and not LILCO's ratepayers, in its lawsuit of LILCO under the Racketeer Influenced and Corrupt Organization law. It is recognized, however, that ratepayers could subsequently collect damages if the suit is successful.

September 15: The Public Service Commission votes 6-1 to accept the Shoreham settlement negotiated by the State and LILCO.

September 23: An Atomic Safety Licensing Board (ASLB), in a 2-1 decision, authorizes the Nuclear Regulatory Commission to issue an operating license for Shoreham, and throws New York State, Suffolk County, and the town of Southampton out as intervenors in the case.

October: U.S. Secretary of Energy John Herrington writes to NRC Chairman Lando Zech asking the NRC to "move expeditiously" to license Shoreham.

October 3: The trial begins in Suffolk County's lawsuit of LILCO under the Racketeer Influenced and Corrupt Organizations Law.

October 7: A Nuclear Regulatory Commission Appeals Board overturns the September 23 decision of the ASLB, saying that the ASLB did not have the right to dismiss the State and County from the entire Shoreham proceeding.

October 30: A *Newsday* poll reports that 50 percent of Long Islanders oppose the State-LILCO settlement, and 31 percent favor it, largely because of the rate increases granted to LILCO.

November 4: LILCO's shareholders vote overwhelmingly to accept the settlement. LILCO Chairman Catacosinos gives the Legislature until December 1 to approve the settlement.

November 7: Governor Cuomo's political committee reports its own poll, concluding that 49 percent of Long Islanders favor the settlement and 26 percent oppose it. The question in the committee's poll includes the statement that "The New York State Public Service Commission has determined that the electric rates if the settlement is approved would be less than if the settlement is not approved and the Shoreham nuclear power plant is allowed to operate."

November 18: President Ronald Reagan issues an executive order specifying that the Federal government itself will provide emergency reception and medical centers, communication services, and other facilities if state and local governments fail to do so in an emergency at a nuclear plant.

November 29: In Suffolk County's RICO suit of LILCO, Judge Jack Weinstein suggests that the County's arguments and evidence are weak, but that he would not take the case away from the jury.

December 1: Legislature fails to approve the settlement by the December 1 deadline imposed by LILCO. However, it is recognized generally that negotiations may continue.

December 5: In the RICO lawsuit of LILCO, a six-person jury finds LILCO guilty of lying to Public Service Commission about Shoreham's progress in order to obtain rate increases.

December 16: Judge Jack Weinstein suggests that LILCO, Suffolk County, and the ratepayers settle the RICO lawsuit prior to his final action in the case or any appeals.

December 29: The New York Public Service Commission votes 6-1 to not grant any additional rate increases to LILCO until the basic terms of the original settlement are agreed to by the key parties, including LILCO, the Public Service Commission, the Long Island Power Authority, and the New York Power Authority.

1989

January 22: Kenneth Feinberg, a mediator proposed by Judge Weinstein, proposes outlines of a settlement of the RICO lawsuit to Suffolk County, LILCO, and attorney appointed by Judge Weinstein to represent ratepayers.

February 11: Judge Weinstein dismisses Suffolk County's lawsuit of LILCO under the Racketeer Influenced and Corrupt Organization law, concluding that the RICO law "cannot, and should not, be applied in a case such as this to permit a federal jury in a civil case to second guess the rate-making authority of the state." Although dismissing the case, Weinstein urges that the parties settle the lawsuit before any further appeals are made.

February 11: Kenneth Feinberg proposes a RICO settlement to Suffolk County, LILCO, and ratepayers, providing $390 million in rate relief over ten years.

February 14: Judith Vladeck, attorney representing ratepayers, approves the settlement proposed by Feinberg.

February 15: Public Service Commission grants LILCO a rate increase, but conditions the increase on LILCO agreeing to not operate Shoreham even if granted an operating license. The PSC approves the increase 7-0, but the conditions for the increase only by a vote of 4-3.

February 27: Suffolk County legislature votes unanimously to appeal Judge Weinstein's dismissal of the RICO case, and refuses to settle the RICO lawsuit with LILCO.

February 28: Governor Cuomo announces that he and the Long Island Lighting Company have agreed on a new settlement to close Shoreham. Under the new settlement LILCO will sell Shoreham to the Long Island Power Authority for $1, which will contract with the New York Power Authority for Shoreham's decommissioning. LILCO obtains a commitment from the State that it will make good faith efforts to return the utility to financial health. The new settlement will not require Legislative approval, and will leave final approval of rate increases solely to Public the Service Commission.

March 3: The Nuclear Regulatory Commission throws out Suffolk County and New York State as intervenors in the Shoreham proceeding, saying that the governments had acted in "bad faith" in "a willful attempt to obstruct the Commission's proceedings."

March 22: Judge Weinstein approves final settlement of the RICO suit for LILCO and its ratepayers, providing $390 million in rate relief, $10 million in attorneys' fees, and funding for a Citizens Advisory Panel to oversee implementation of the agreement. Suffolk County's legislators reaffirm their opposition to the settlement of the RICO suit, and announce they will press an appeal of Weinstein's dismissal of the case.

April 5: New York Public Service Commission approves new State-LILCO settlement by a vote of 6-1, and approves a ten-year plan for annual rate increases of between four and five percent for LILCO.

April 12: New York Power Authority votes 3-2 to arrange decommissioning for Shoreham and provide new power plants on Long Island if asked to do so as provided under the settlement.

April 12: James Watkins, the United States Secretary of Energy, calls the New York-LILCO settlement of Shoreham "one of the most foolish deals in the nation's history." Deputy Energy Secretary Henson Moore says that Department of Energy will "throw up every roadblock we can to it, and if we have to create some, we'll do that."

April 13: The Long Island Power Authority votes 8-1 to approve the new settlement.

April 13: United States Senate Committee on Energy and Natural Resources holds hearing on electricity supply in the Northeast which is generally critical of the New York-LILCO settlement.

April 14: LILCO Board of Directors votes unanimously to approve the new settlement.

April 17: Nuclear Regulatory Commission staff issues formal finding that Shoreham is acceptably safe and that Long Island's geography or other features do not "render emergency planning at Shoreham fundamentally more difficult than at other sites."

April 20: Nuclear Regulatory Commission votes to grant LILCO full commercial operating license for Shoreham.

June 28: LILCO's stockholders vote overwhelmingly to approve the New York-LILCO settlement leading to abandonment of Shoreham in exchange for a package of financial aid from the State..

June 29: U.S. House Committee on Energy and Commerce passes, 23-19, an amendment to NRC authorization bill sponsored by Representative Donald Ritter (R-PA) which prohibits the NRC from transferring LILCO's Shoreham license to the Long Island Power Authority.

July: U.S. Senate Committee on Energy and Natural Resources releases study by staff economist Karl Hausker criticizing New York's economic evaluation of settlement as being based on a "deeply flawed...obviously biased" study by the New York State Public Service Commission.

July 18: House Speaker Thomas Foley delays full House vote on NRC authorization bill including the Ritter amendment.

August 15: President Bush, Deputy Energy Secretary Henson Moore, and Governor Cuomo meet on the Shoreham issue.

September 1: Federal government requests to join a suit in New York State which would block implementation of the settlement.

November 14: President Bush affirms administration policy that it would be better to operate Shoreham than to decommission the plant.

November 16: House debates NRC authorization including Ritter amendment. House votes 405-11 to delay final consideration of bill after Interior Committee Chairman Morris Udall says that Ritter amendment might constitute Legislative "taking" of LILCO's property, requiring Congress to compensate LILCO for the full cost of the plant.

1990

January 3: Governor Mario Cuomo's State of the State address calls for the possible conversion of Shoreham to a gas-fired facility.

Chapter 8

CONCLUSION

This Chapter makes some concluding comments about the Long Island Lighting Company, State and local policy, and what seemed to be underlying dynamics in Shoreham's history. Similar dynamics might be found in other disputes over complex, expensive, and potentially dangerous technologies.

LILCO'S Problems and Changes

The Long Island Lighting Company largely managed itself into the Shoreham crisis by mishandling the plant's construction, thereby losing legitimacy, time, and money. Combined with LILCO's other problems, such as fluctuations in the demand for electric power, its high prices, and its already negative image, Shoreham threatened to kill the company in the 1980s.

LILCO also, I would argue, managed its way *out* of the Shoreham crisis. First, and most obviously, LILCO negotiated with New York State a Shoreham settlement designed to return the company to an investment grade rating reasonably quickly. This settlement, by any reasonable standard, favored the company. Second, and perhaps more striking, by the late 1980s LILCO had developed the capabilities to manage a nuclear power plant--capabilities the company had not shown during Shoreham's construction. The political environment in New York, however, was unforgiving. By that time, the political and personal commitments against Shoreham were so strong that the State government would not compromise its opposition to the plant. While LILCO could have operated Shoreham effectively, under the circumstances the company was quite willing to "sacrifice" it on favorable financial terms. This Section discusses these issues.

LILCO Managed Its Way Into Much of the Crisis

LILCO managed Shoreham's construction ineffectively, and was vulnerable when State and local governments joined the citizens' movement against the plant in the early 1980s. Other nuclear projects shared many of these management problems (Cook, 1985; Leigland and Lamb, 1986; Tomain, 1987: chapter 2).

Prior to Shoreham, LILCO was regarded as a managerially strong and experienced company. It was solid financially, as demonstrated by long-term debt ratings and other indicators. Moody's Investor Service rates corporate bonds on investment value on a nine-grade scale: **Aaa, Aa, A, Baa, Ba, B, Caa, Ca,** and **C**, with **Aaa** the highest level and **C** the lowest. Preferred stocks are rated on a nine-grade scale of **aaa** through **c**. Table 18 presents LILCO's ratings, along with its common stock dividends, for the period 1966 through 1988. The Table indicates that, until the 1980s, the company's position was sound.

Similarly, the New York State Public Service Commission, under normal circumstances, would not intervene strongly in the management of the company, for earlier LILCO had built and operated fossil fuel plants successfully. However, gradually it became clear that construction of a nuclear plant required a high level of construction management skills. As discussed in Chapter 1, nuclear power is technologically more complex than power from fossil fuels. The nuclear plants under construction in the 1970s tended to be larger than fossil fuel plants, and they also were held to more exacting standards because of the safety issue. An official from the Nuclear Regulatory Commission commented that Shoreham was held to particularly high standards because the NRC expected, on the basis of the lengthy construction permit hearings from 1970-1973, that Shoreham's operating license would be contested strongly (U.S. Office of Technology Assessment, 1984: 118-121; New York State Public Service Commission, 1985: 85-86).

During the 1970s, LILCO lacked the experience, awareness, and skills required to manage the Shoreham project. As discussed in Chapter 4, the Public Service Commission's prudence investigation demonstrated that during construction LILCO failed in planning and controlling the project, dealt with the Nuclear Regulatory Commission in an ineffective way, and managed its relationship with Stone and Webster poorly. The PSC ultimately disallowed about one-third of Shoreham's costs to 1984 as being incurred imprudently.

Table 18. LILCO's Bond and Stock Ratings and Stock Dividends

	Ratings[1]		
	Bonds[2]	Preferred Stock[3]	Common Stock Dividends
1966	Aa		$1.06
1967	Aa		1.14
1968	Aa		1.22
1969	Aa		1.28
1970	Aa		1.33
1971	Aa		1.37
1972	Aa		1.41
1973	Aa		1.45
1974	Aa		1.46
1975	Aa	a	1.49
1976	Aa - A	a	1.55
1977	Aa - A	a	1.63
1978	Aa - A	a	1.70
1979	A	a	1.76
1980	A - Baa	baa	1.84
1981	A - Baa	baa	1.92
1982	Baa1 - Baa2	baa3	2.00
1983	Baa2 - Baa3	ba1	2.02
1984	B1 - B2	caa	.0
1985	Ba3 - B1	caa	.0
1986	Ba3 - B1	caa	.0
1987	Ba3 - B1	caa	.0

1. Beginning in 1982 Moody's attached a 1 to 3 numerical modifier to grades, indicating the upper or lower tier of the grade.

2. In the period LILCO's bond ratings were bounded by the Aa and B1 grades. Aa grades mean that bonds "are judged to be of high quality by all standards," although somewhat less protected than those with the very highest ratings. B1 grades "lack characteristics of a desirable investment."

3. LILCO's grades were bounded by the a and caa ratings. An a grade indicates that a stock's "earnings and asset protections are...expected to be maintained at adequate levels." The caa grade signifies that the stock "is likely to be in arrears on dividend payments."

Source: *Moody's Public Utility Manual*, various years.

Electricity Demand Softened As Shoreham Neared Completion

The Shoreham problem interacted with LILCO's other difficulties. These included fluctuations in demand for electricity and, in particular, demand which frequently declined on an annual basis as Shoreham neared completion. Under the circumstances it could be argued that the controversial plant's electricity would not be needed.

The demand for electricity slowed substantially nationwide during the 1970s (Table 19). From 1965 to 1974, when large numbers of nuclear plants were ordered, electricity consumption nationally had increased 79 percent. In contrast, from 1975 to 1984 electricity consumption increased 31 percent (U.S. Department of Energy, 1988b: 191). Indeed, for several years on Long Island between 1974 and 1984 peak demand actually fell. Many utilities cut back plans to expand generating capacity, and nuclear facilities were most likely to be cut because of their political and economic riskiness. The Long Island Lighting Company moved ahead with Shoreham, but as the plant neared completion Shoreham's opponents argued repeatedly that its electricity simply was not needed. The demand for electricity surged in the late 1980s, but by that time Governor Cuomo had committed New York State to blocking the plant's commercial operation.

LILCO's High Electricity Rates Damaged Its Credibility

The Long Island Lighting Company had the second highest electricity rates in the nation (Consolidated Edison in New York City was first), and its rates had increased substantially during the Shoreham controversy (Table 20). A relatively small portion of Shoreham's costs had been added to LILCO's rate base in the 1970s; the price increases primarily reflected other factors. (This meant that still greater increases would come when the plant either opened or was settled by LILCO and the State.) For example, LILCO was almost totally dependent on imported oil for fuel. From 1977 to 1982 the price of distillate fuel for electric utilities, *in constant dollars*, increased 73 percent, and the price of residual fuel, also in constant dollars, increased 49 percent (U.S. Department of Energy, State Energy Price and Expenditure Report 1970-1982, p. 17).

Nationally, from 1970 to 1987, weighted average monthly electric bills increased about 42 percent in constant 1982 dollars (U.S. Department of Energy Annual Energy Review, 1987: 207). For the same period, in constant dollars, LILCO's revenue per kilowatt hour increased about 90 percent for industrial customers and 80 percent for residential customers. Between 1976 and 1981, the five years leading up to the withdrawal of County and State support, average national electric bills increased 10

Table 19. Percentage Change in Indicators of Electricity Generation

Year	United States[1]	New York State[2]	LILCO[3]
1971	5.3%	1.8%	10.5%
1972	8.5	4.7	9.2
1973	6.3	4.1	11.5
1974	.3	1.3	-4.4
1975	2.7	-.8	4.7
1976	6.3	3.1	-7.3
1977	4.2	2.2	14.3
1978	3.9	1.6	-3.5
1979	1.9	2.0	-2.6
1980	1.7	1.1	7.7
1981	.4	-.9	-.3
1982	-2.3	-.4	-2.8
1983	3.1	3.7	2.1
1984	4.6	2.3	-.4
1985	2.2	2.1	9.2
1986	.7	2.0	1.8
1987	3.4	3.8	3.9
1988	5.0	--	6.9

1. Percentage change in United States consumption of kilowatt hours from electric utilities. Source: U.S. Department of Energy, *Annual Energy Review 1988* (DOE/EIA-0384(88) 1989 (p. 197)

2. Percentage change in kilowatt hours generated in New York State (includes net imports of electricity). *New York State Annual Energy Review: Consumption, Supply, and Price Statistics 1970-1987*. New York State Energy Office, 1988: 14.

3. Percentage change in peak demand. Source: Long Island Lighting Company Annual Reports.

Table 20. Revenue Per Kilowatt Hour, Long Island Lighting Company (Cents per kilowatt hour)

	Industrial Customers		**Residential Customers**	
Year	Current $	Constant $ (1982)	Current $	Constant $ (1982)
1970	1.99	4.73	2.42	5.76
1971	2.24	5.04	2.67	6.01
1972	2.35	5.05	2.79	6.00
1973	2.52	5.09	2.99	6.04
1974	3.97	7.35	4.48	8.30
1975	4.46	7.52	4.99	8.41
1976	4.58	7.26	5.19	8.22
1977	5.16	7.67	5.80	8.62
1978	5.39	7.46	6.27	8.68
1979	6.25	7.95	7.16	9.11
1980	7.46	8.70	8.46	9.87
1981	10.26	10.91	11.37	12.10
1982	9.80	9.80	10.95	10.95
1983	10.21	9.83	11.43	11.00
1984	11.17	10.37	12.53	11.63
1985	11.46	10.30	12.95	11.65
1986	10.43	9.14	11.92	10.44
1987	10.62	9.04	12.13	10.32

Source: Long Island Lighting Company Annual Reports. GNP Price Deflator from United States Department of Energy, *Annual Energy Review 1987*, (DOE/EIA-0384(87) May 1988), p. 283.

percent, while LILCO's revenues in constant dollars per kilowatt hour increased around 50 and 47 percent for industrial and residential customers, respectively.

LILCO's reputation as one of the most expensive electrical systems in the nation contributed to its problems from the general controversy over nuclear power, its mismanagement of Shoreham's construction, and the effects of unstable demand for electricity. Shoreham's opponents portrayed LILCO as a managerially inept company charging grotesque prices for electricity, while trying to force through a dangerous nuclear plant even as the need for electricity was falling. This image mixed political rhetoric and fact. It was true that the Public Service Commission concluded that LILCO had mismanaged Shoreham's construction; that LILCO's prices were the second highest in the nation; that Shoreham's safety was disputed by a strong anti-nuclear movement on Long Island and many individuals in key elected positions; and that the demand for electricity had softened as Shoreham neared completion. LILCO's opponents crafted these facts into a politically powerful image of an irresponsible electric utility, and certainly not a company to be trusted operating dangerous technology.

The Movement to Eliminate LILCO

Shoreham's construction permit hearings from 1970-1973 were the longest and most divisive in the Atomic Energy Commission's history. The Three Mile Island accident in 1979, and the imminent completion of the plant in the early 1980s, brought new life to the anti-Shoreham movement. By that time the issue was no longer just Shoreham, but also excising from Long Island the corporation that had created it. The active movement against nuclear power tends to mistrust large corporations in general (Douglas and Wildavsky, 1982; Soderstrom, Sorensen, Copenhaver, and Carnes, 1984; Buss, Craik, and Dake, 1986). LILCO's problems left it particularly open to comprehensive attack, and to swings against it by broader public opinion. In 1983 two members of the State Fact Finding Panel on Shoreham wrote that

> The panel heard much testimony about the consequences of high [electricity] rates, but focused rather too infrequently on proposals for moderating or even reducing rates, clearly desirable for the area's economic vitality. As the rate performance of non-profit, municipally or citizen-owned utilities is enviable in comparison with private utility companies, it is not surprising that public power was raised as a public policy for serious consideration...Long Islanders..., awakened by LILCO's trespasses, seek locally accountable, elected utility boards...There is also wide agreement among public power advocates, both on LI and in Albany, that a Long Island public power system should not be a nuclear owner and/or operator (Willmott and Harrison, 1983: 58, 60).

In 1988 anti-Shoreham forces divided on whether the State should negotiate a Shoreham settlement with LILCO, or wholeheartedly support a public takeover. Long Island Power Authority member Irving Like argued that "If LILCO remains in existence its hegemony will continue. All of its connections and all of its media hype, all of that dynamic will continue and will seek to undermine a settlement and open Shoreham" (*New York Times*, February 29, 1988: II-3). Governor Cuomo characterized some of the opponents of the negotiated settlement as "the extreme left, which is for a public takeover, even risking the opening of Shoreham" (*Newsday*, July 5, 1988). Finally, Judge Jack Weinstein, in affirming the negotiated LILCO-ratepayer settlement of the RICO suit in 1989, wrote that

> This case...brings to the surface private agendas for publicly owned power and deep antipathy towards large corporations, particularly utilities. It has elements of xenophobia and hysteria. It reflects a deep populist strain of our citizens and an urge to stand up for what they believe in, to speak out on public issues and to be heard by public officials. Stronger than the other rational and emotional aspects of the case is a pervasive fear of atomic energy generally and the Shoreham Nuclear Power Station (Shoreham) in particular; if there was one overriding theme of plaintiffs in the case it was that Shoreham must close. To understand this case is to appreciate that money alone and economic theory of litigation do not fully explain the dynamics of mass tort litigation (County of Suffolk et al. v. Long Island Lighting Company et. al., United States District Court, Second Circuit Eastern District, 88-CV-2065 (JBW) [March 22, 1989]).

LILCO's Organizational Performance Improved In the Mid-1980s

As discussed in Chapter 1, utilities' performance with nuclear plants seemed to improve with experience. Utilities with a higher share of their investment in nuclear plants appear to have better operating reliability records (Osborn and Jackson, 1988), and utilities with substantial nuclear experience apparently have better construction records (Lester, 1986). The Long Island Lighting Company's performance through Shoreham's history is consistent with the theme of organizational learning.

LILCO's political and financial situation in the mid-1980s was poor both because of its own ineffectiveness in key areas and external pressures. But crises sometimes prompt changes which strengthen an organization. Crises make obvious the need for organizational changes, and make the changes more feasible; the internal level of commitment might rise because of external threats; tolerance of errors decreases; and executives, linked substantively and/or symbolically to past failures, are replaced. LILCO's corporate performance did appear to improve in a number of respects beyond the mid-1980s.

As Chapter 4 discussed, the Nuclear Regulatory Commission strongly criticized LILCO's management of Shoreham's construction during the 1970s and early 1980s (See chapter 4). But in 1986 and 1987 the NRC gave LILCO relatively favorable performance reviews. The NRC's Systematic Assessment of Licensee Performance for LILCO for 1987 stated that "the Shoreham nuclear power station staff and management has maintained the high performance levels for areas previously assessed as strengths and effected improvements in other areas," and that in many cases improvement was due to to increased involvement of senior managers at the plant (*Newsday*, December 4, 1987). Table 21 gives the ratings for the 1986 and 1987 assessments. (Three ratings are given in the review: "excellent," "satisfactory," and "needs attention."). In 1988, Clay Warren, the NRC's resident inspector for the Shoreham facility, attributed improved performance in quality assurance and in relations with the NRC to management changes at the company. He commented that "When I have presented them with the concerns, they have been very responsive in answering those concerns" (Fitzgerald, 1988: 34).

Similarly, in 1986 LILCO initiated several programs to improve service reliability, including a 24-hour telephone service to register and respond to customer problems. In 1982 the State Public Service Commission began recording complaints made to it about utility service. Table 22 indicates the annual complaint rates for the electric and gas utilities in New York State with more than one million customers in 1989. Complaint rates for each of the utilities fell between 1982--when the information system was established--and 1988. LILCO's complaints rose in 1984-1985, and then fell sharply. In 1985 only two of the ten utilities had complaint rates higher than LILCO's. In contrast, for 1988 LILCO had the lowest complaint rate for the four large utilities above, and the third lowest rate among the ten listed regulated electric/gas utilities in the State. An analysis prepared within the Public Service Commission concluded that the drop in LILCO's complaints was due heavily to its new service program (Sawyer, undated). LILCO had strong incentives to improve its reputation with the Public Service Commission and customers, particularly given the precariousness of its situation at the time.

To summarize, organizations are capable of learning and improving their performance over time. By the mid- to late-1980s the Long Island Lighting Company appeared able to operate Shoreham effectively, and had strengthened its management more generally. Several reviews, by different Atomic Safety Licensing Boards, had resolved virtually all of the technical issues regarding Shoreham's management in the company's favor. The Nuclear Regulatory Commission, which had criticized the company strongly earlier, granted LILCO an operating license in 1989. It would have been rational, technically, for New York State to reassess its policy toward Shoreham given the favorable external evaluations of the plant and the changes at LILCO. If anything, however, the State's policy toward blocking the plant became more intransigent.

Table 21. Nuclear Regulatory Commission Assessment of LILCO's Performance, 1985-1987

	Assessment Period	
Area of Performance	March 1, 1985- Feb. 28, 1986	March 1, 1986- July 31, 1987
Plant operations and start-up testing	Satisfactory	Excellent
Radiological controls	Needs attention	Excellent
Maintenance	Satisfactory	Satisfactory*
Surveillance	Satisfactory	Satisfactory
Engineering and corporate technical support	Not evaluated	Satisfactory
Emergency preparedness	Excellent	Excellent
Security and safeguards	Excellent	Excellent
Training and qualification effectiveness	Needs attention	Satisfactory*
Licensing activities	Needs attention	Excellent
Assurance of quality	Satisfactory	Excellent
Outage management, modifications, and technical support	Satisfactory	Not evaluated

*Trend of these areas was assessed as improving

Source: Nuclear Regulatory Commission, as reported in *Newsday*, December 4, 1987.

State and Local Opposition

Until the early 1980s the Suffolk County government had supported or at least tolerated the Shoreham project. In 1982 the County government reversed its position and vowed to block the plant by whatever means necessary; refusing to participate in the emergency planning required for an operating license was the major tactic.

By 1979 the New York State government effectively had imposed a moratorium on the construction of new nuclear power plants, but had not opposed Shoreham's imminent completion and operation. Several other nuclear power plants were already operating in New York State, and so the idea of finishing a largely constructed plant was acceptable.

But in 1983 Governor Mario Cuomo, who had earlier expressed reservations about the use of nuclear power, directed the State to not replace Suffolk County in the emergency planning process. Subsequently, the Governor committed the State government as an active intervenor to block Shoreham. The Governor identified himself so closely with the movement against Shoreham and LILCO, and portrayed the issue in such starkly moralistic terms, that compromise on the issue of operating Shoreham seemed to be out of the question politically.

Governor Cuomo's mobilization of the State's executive branch weighed heavily against Shoreham. The Governor could have directed the State to replace Suffolk County in emergency planning--as he had directed the State to replace Rockland County in emergency plan exercises at the Indian Point nuclear plant in 1983--but did not. His support was central to the formation of the Long Island Power Authority, and to the early credibility of the threat to replace LILCO with a system of "public power" on Long Island. The Governor was able to reinforce public hostility toward LILCO by speaking out aggressively against the company, further weakening its political position (Declaring at one point, for example, that "LILCO rips off the people of Long Island. Now we are coming after you."). By reconstituting the Public Service Commission in 1987 he could be fairly confident that LILCO would receive steady financial pressure to resolve the controversy on terms which would be acceptable to the State; subsequently, the Public Service Commission linked its decisions on LILCO's rate increases to the company's progress in abandoning the plant. Governor Cuomo also reinforced public fears about nuclear power, repeating charges that effective emergency planning for Shoreham was impossible, that operating the plant at 25 percent of power was "one quarter of the way to devastation," and so forth.

Table 22. Complaints Made To Public Service Commission (Per 100,000 Customers)

	Consolidated Edison	LILCO	Niagara Mohawk	Brooklyn Union Gas
1982	88.1	31.0	24.9	72.1
1983	103.3	30.7	20.7	70.0
1984	104.5	38.8	15.4	54.8
1985	70.7	37.5	13.2	41.1
1986	44.7	22.4	10.7	22.8
1987	27.9	9.7	9.7	21.2
1988	20.9	6.0	9.5	14.7

Source: Data provided by New York State Public Service Commission, Office of Consumer Affairs (June, 1989)

Finally, to close Shoreham the Governor was willing to compensate LILCO for most of Shoreham's prudently incurred costs, and otherwise aid the company in its life after Shoreham. In fact, one of the difficult balancing acts the Governor faced was, at one moment, instigating a feeding frenzy against LILCO on Long Island, and, at the next, supporting an agreement which included a generous package of financial assistance to LILCO.

Policy, Personal Stakes, and Decision Momentum

Writing of one particular dispute over Shoreham, reporter Gary Spencer commented that the dispute "may well have been inevitable, stemming as it does from conflicts of policy, politics, and ego" (*Newsday*, November 27, 1988). Although restricted to one issue, Spencer's comment effectively summarizes Shoreham's history. Nuclear power involves difficult *policy* choices concerning amounts, types, and prices of energy, regulation of health and environmental issues, and other important matters. These choices take place in a *politically* complex and contentious environment. But individuals' *personal* beliefs, stakes and commitments, as well, made a great deal of difference in the outcome at Shoreham.

The three factors became mingled over the period discussed here. In many respects, the underlying dynamics of the story seem familiar. Studies of implementation suggest that the longer a project takes and the more actors it involves, the greater the difficulties of bringing it "on line." The early enthusiasm and support which generated the project in the first place might dull; or, the tolerance a project receives initially may decay. Major projects almost by definition redistribute resources, change economic and political situations, and favor the values of some groups over others. As time goes on the actors who are relatively disadvantaged by a project--either because they lose resources or see their values threatened--discover the extent of their losses and try to block it. These actors generally are able to take advantage of the relatively open system of administrative law and politics in the United States. Neither side may be able to "win" a dispute quickly. The supporters are able to make some progress in getting a project completed and approved; however, opponents are able to throw up obstacles which delay the project and raise its costs (Stewart, 1975; Thurow, 1980; Pressman and Wildavsky, 1979).

In a long-running controversy particular actors on both sides live with the issue through successful and unsuccessful skirmishes. Gradually the symbolic and personal stakes of individuals in the conflict fuel intense and emotional commitments to "winning." Under the circumstances efficient compromises are unlikely to be made (Douglas and Wildavsky, 1984). It is likely that the Long Island Lighting Company eventually was able to compromise its goal of operating the plant because, by the late 1980s, the executives and managers who had built and defended Shoreham so vigorously had been replaced with others who were not identified so closely with the project. The new executives were content to escape from the controversy with the company intact financially.

In contrast, the actors in Suffolk County and the Governor's office who started with the controversy early in the decade (or even earlier) remained with it through to the finish. Their longstanding policy, political, and personal commitments were to stop the Shoreham nuclear plant. When LILCO moved into a position to receive an operating license, they, as public decision makers, ultimately were willing to pay a great deal--about $750 million on a present value basis, according to the 1989 Senate study--to purchase a future without a nuclear plant at Shoreham (U.S. Senate Committee on Energy and Natural Resources, 1989b).

Thus, the State, County, and LILCO compromised on Shoreham, but--absent information that the plant was particularly risky--the compromise was not efficient. There is a rich public record of how many utilities managed themselves into organizational disasters in nuclear projects. In Shoreham's history, the Long Island Lighting Company managed itself into a crisis, but showed an ability over the lifetime of the issue to learn and adapt. The State and local government failures, however, were more decisive and striking.

REFERENCES

Ahearne, John. 1987. "Nuclear Power After Chernobyl." *Science* 236: 673-679 (May 8).

Arthur D. Little, Inc. 1977. *Management Audit of Long Island Lighting Company.* Boston: Arthur D. Little, Inc.

_____. 1984. *Management Audit of Long Island Lighting Company.* Boston: Arthur D. Little, Inc.

Asset Transfer Agreement Between Long Island Lighting Company and Long Island Power Authority. June 15, 1988. Obtained from New York State Public Service Commission.

Bupp, I. C., and J. Derian. 1978. *Light Water: How the Nuclear Dream Dissolved.* New York: Basic Books.

Buss, D., K. Craik, and K. Dake. 1986. "Contemporary Worldviews and Perception of the Technological System." In V. Covello, J. Menkes, and J. Mumpowers, eds., *Risk Evaluation and Management*, pp. 93-130. New York: Plenum Publishing Corporation.

Campbell, J. L. 1988. *Collapse of an Industry.* Ithaca: Cornell University Press.

Carl, A. 1972. "The Lloyd Harbor Study Group Intervention--A Response." *Science and Public Affairs* XXVIII (6): 31-36.

Clark, L. 1985. "The Origins of Nuclear Power: A Case of Institutional Conflict." *Social Problems* 32: 474-487 (June).

Cook, J. 1985. "Nuclear Follies." *Forbes*, February 11: 82-100.

Del Sesto, S., 1979. *Science, Politics, and Controversy: Civilian Nuclear Power in the United States, 1946-74.* Boulder, Colorado: Westview Press.

Diamond, S. November 1981. *Shoreham: What Went Wrong.* Long Island, New York: *Newsday* Reprint.

Douglas, M., and A. Wildavsky. 1982. *Risk and Culture.* Berkeley: University of California Press.

Energy Systems Research Group, Inc. 1988. Testimony of Neil Talbot Concerning Proposed LILCO Settlement (July 20). Albany, New York: Public Service Commission.

Fitzgerald, K. 1987. "Regulation and Retrenchment." *IEEE Spectrum* 24 (November): 30-34.

Fitzgerald, K., and G. Zorpette. 1987. "The First Decade: Great Expectations," *IEEE Spectrum* 24 (November): 25-30.

Ford, D. 1982. *Cult of the Atom: The Secret Papers of the Atomic Energy Commission*. New York: Simon and Schuster.

Gamson, W. A. and A. Modigliani. 1989. "Media Discourse and Public Opinion on Nuclear Power: A Constructionist Approach." *American Journal of Sociology*, 95: 1-37 (July).

Green, H. 1972. "Nuclear Safety and the Public Interest." *Nuclear News* 15 (September): 75-78.

Helm, C. 1987. "Is Opposition to Nuclear Energy an Ideological Critique?" *American Political Science Review*, 82: 943-946 (September).

Herr P. B. and Associates. 1982. *Discussion Overview of the Radiological Emergency Response Plan of the County of Suffolk*. In U.S. Senate Committee on Environment and Public Works, *Hearings on Nuclear Emergency Planning*: 343-365. Washington: U.S. Government Printing Office.

Komanoff, C. 1981. *Power Plant Cost Escalation: Nuclear and Coal Capital Costs, Regulation, and Economics*. New York: Van Nostrand Reinhold Company.

Kouts, H. 1983. "A History of Nuclear Power Development." In *Final Report of New York State Fact Finding Panel on the Shoreham Nuclear Power Facility*, Appendix 5(b). Stony Brook, New York: New York State Fact Finding Panel.

Lazard Freres and Company. 1988a. *Comparative Rate Analysis of Public Acquisition of Long Island Lighting Company* (Volumes I and II). New York: Lazard Freres and Company.

____. 1988b. *Comparative Rate Analysis: Settlement Scenario vs. LIPA Scenario* (June 21). New York: Lazard Freres and Company.

_____. 1988c. *Comparative Rate Analysis: Settlement Scenario vs. LILCO-Shoreham In Scenario* (June 21). New York: Lazard Freres and Company.

Leigland, J., and R. Lamb. 1986. *WPP$$: Who Is to Blame for the WPPSS Disaster*? Cambridge: Ballinger Publishing Company.

Lester, R. 1986. "Organization, Structure, and Performance in the U.S. Nuclear Power Industry." *Energy Systems and Policy* 9: 335-384.

Long Island Lighting Company. 1984. *Suffolk County Takeover of LILCO: Risks and Consequences* (November). Long Island, New York: Long Island Lighting Company.

Marcus, A. 1986. "Scientific Uncertainty and Shifts in the Policy Subsystem for Nuclear Power." Working Paper, University of Minnesota School of Management (January).

_____. 1988. "Implementing Externally Induced Innovations: A Comparison of Rule-bound and Autonomous Approaches." *Academy of Management Journal* 31: 235-256.

Mazuzan, G., and J. S. Walker. 1984. *Controlling the Atom*. Berkeley: University of California Press.

Memorandum of Understanding Concerning Proposed Agreements on Power Supply for Long Island Between Long Island Lighting Company and New York Power Authority. 1988 (undated). Obtained from State of New York Public Service Commission.

Morone, J., and E. Woodhouse. 1989. *The Demise of Nuclear Energy?* New Haven: Yale University Press.

National Economic Research Associates, Inc. (NERA). 1986. *Will A Government Takeover of LILCO Save Money for Consumers?* White Plains, New York: National Economic Research Associates, Inc.

Navarro, P. 1985. *The Dimming of America*. Cambridge: Ballinger Publishing Company.

New York State Consumer Protection Board. 1988. Testimony of Richard Kessel in Support of LILCO Settlement (July 1). Albany, New York: Public Service Commission.

New York State Department of Public Service Staff. 1984. *Long Island Lighting Company Case 27563: Staff's Initial Brief*. Albany, New York: State of New York Public Service Commission.

_____. 1988a. Memorandum on Proposed Settlement on Shoreham and LILCO (June 23). Albany, New York: State of New York Public Service Commission.

_____. 1988b. Testimony Concerning Energy Systems Associates Estimates. Albany, New York: State of New York Public Service Commission.

New York State Fact Finding Panel on the Shoreham Nuclear Power Facility. 1983. *Final Report* (December). Stony Brook, New York: New York State Fact Finding Panel.

New York State Fact Finding Panel on the Shoreham Nuclear Power Facility Transcript. 1983. Meeting of October 1, 1983. Stony Brook, New York: New York State Fact Finding Panel.

New York State Fact Finding Panel on the Shoreham Nuclear Power Facility Staff. 1983. "The Nature and Manner of Risks Associated with the Operation of the Shoreham Nuclear Power Plant." In *Final Report* of Fact Finding Panel, pp. 3(a)1-57). Stony Brook, New York: New York State Fact Finding Panel (December).

New York State Long Island Public Power Panel. 1986. *Final Report* (June 24). Albany, New York: Long Island Public Power Panel.

New York State Public Service Commission. 1985. *Long Island Lighting Company Case 27563: Opinion and Order Determining Prudent Costs*. Albany, New York: State of New York Public Service Commission.

_____. 1989. *Comparative Economics of Shoreham Settlement and Shoreham In-Service* (April 14). Albany, New York: State of New York Public Service Commission.

Osborn, R. and D. Jackson. 1988. "Leaders, Riverboat Gamblers, or Purposeful Unintended Consequences in the Management of Complex, Dangerous Technologies." *Academy of Management Journal,* 31: 924-947.

Perrow, Charles. 1984. *Normal Accidents: Living With High-Risk Technologies*. New York: Basic Books.

Pressman, J. and A. Wildavsky. 1979. *Implementation*. Berkeley: University of California Press.

Robinson, F. 1988. Memorandum to New York State Public Service Commission on Analysis of Shoreham Settlement (Cases 29484 and 88-E-084 (September 1). Albany: New York State Public Service Commission.

Rothman, S., and S. R. Lichter. 1987. "Elite Ideology and Risk Perception in Nuclear Energy Policy." *American Political Science Review*, 81: 383-404 (June).

_____. 1988. Response to "Is Opposition to Nuclear Energy An Ideological Critique." *American Political Science Review*, 82: 947-951.

Sailor, V. 1972. "The Role of the Lloyd Harbor Study Group in the Shoreham Hearings--An Assessment." *Science and Public Affairs* XXVIII (6): 25-31.

Sawyer, David. undated. "Controlling Utility PSC Complaints." Obtained from New York State Public Service Commission Office of Consumer Affairs, June, 1989.

Settlement Agreement - LILCO Issues. February 28, 1989. Obtained from New York State Public Service Commission.

Settlement Agreement - LILCO Issues. June 16, 1988. Obtained from New York State Public Service Commission.

Shattuck, Petra. 1986. "Federalism and Offsite Emergency Planning for Nuclear Reactors: The Shoreham Impasse." *Boston University Law Review* 66 (2): 229-269.

Slovic, P., B. Fischoff, and S. Lichtenstein. 1986. "The Psychometric Study of Risk Perception." In V. Covello, J. Menkes, and J. Mumpowers, eds., *Risk Evaluation and Management*, pp. 3-24. New York: Plenum Publishing Corporation.

Soderstrom, E. J., J. Sorenson, E. Copenhaver, and S. Carnes. 1984. "Risk Perception in an Interest Group Context: An Examination of the TMI Restart Issue." *Risk Analysis*, 4: 231-244.

Stein, C. 1986. "The Stone and Webster Effect." *Boston Globe*, July 8: 35, 40.

Stewart, R. 1975. "The Reformation of American Administrative Law." *Harvard Law Review* 88: 1669-1813.

Thurow, L. 1980. *The Zero Sum Society*. New York: Basic Books.

Tomaine, J. 1987. *Nuclear Power Transformation*. Bloomington: Indiana University Press.

Union Associates. 1988. *Comment on Proposed Acquisition of Shoreham by the Long Island Power Authority* (June 22). New York: Union Associates.

United States Atomic Energy Commission. 1958. *Theoretical Possibilities and Consequences of Major Accidents in Large Nuclear Plants, WASH-740*. Washington: U.S. Government Printing Office.

United States Atomic Energy Commission. 1967. *Civilian Nuclear Power: The 1967 Supplement to the 1962 Report to the President.* Washington: U.S. Atomic Energy Commission.

United States Congress, House Committee on Energy and Commerce. 1986. *Hearings on FEMA and Nuclear Powerplant Safety*. Washington: U.S. Government Printing Office.

United States Congress, Joint Committee on Atomic Energy. 1968. *Nuclear Power Economics 1962 Through 1967*. Washington: U.S. Government Printing Office.

United States Congress, Senate Committee on Energy and Natural Resources. 1989a. *Hearings on U.S. Electricity Demand--The Northeastern Region*. Washington: U.S. Government Printing Office.

_____. 1989b. *The Economics of the Shoreham Nuclear Plant*. Washington: U.S. Government Printing Office.

United States Congress, Senate Committee on Environment and Public Works. 1983. *Hearings on Nuclear Emergency Planning*. Washington: U.S. Government Printing Office.

_____. 1987a. *Reauthorization of the Nuclear Regulatory Commission for Fiscal Years 1988 and 1989 and Nuclear Emergency Planning*. Washington: U.S. Government Printing Office.

United States Congress, Senate Committee on Environment and Public Works. 1987b. *Hearings on Shoreham Nuclear Powerplant Licensing Issues*. Washington: U.S. Government Printing Office.

United States Department of Energy, Energy Information Administration. 1983. *Nuclear Plant Cancellations: Causes, Costs, and Consequences*. Washington: U.S. Government Printing Office.

_____. 1985. *State Energy Demand and Expenditure Report, 1970-82* (April). Washington: U.S. Government Printing Office (1985a).

_____. 1985. *An Analysis of Nuclear Power Construction Costs*. Washington: U.S. Government Printing Office (1985b).

_____. 1988. *Commercial Nuclear Power 1988: Prospects for the United States and the World*. Washington: U.S. Government Printing Office (1988a).

_____. 1988. *Annual Energy Review, 1987*. Washington: U.S. Government Printing Office (1988b).

_____. 1989. *Annual Energy Review, 1988*. Washington: U.S. Government Printing Office.

United States Federal Emergency Management Agency. 1988. *Post-Exercise Assessment: Shoreham Nuclear Power Station* (September 2). New York: Federal Emergency Management Agency, Region II.

United States General Accounting Office. 1979. *Areas Around Nuclear Facilities Should Be Better Prepared for Radiological Emergencies*. Washington: U.S. General Accounting Office.

United States Nuclear Regulatory Commission and U.S. Environmental Protection Agency Task Force on Emergency Planning. 1978. *Planning Basis for the Development of State and Local Government Radiological Emergency Response Plans in Support of Light Water Nuclear Plants (NUREG-0396)* (December). Washington: Nuclear Regulatory Commission.

United States Nuclear Regulatory Commission Special Inquiry Group. 1979. *Three Mile Island: A Report to the Commissioners and the Public*. Washington: U.S. Nuclear Regulatory Commission.

United States Nuclear Regulatory Commission and Federal Emergency Management Agency. 1980. *Criteria for Preparation and Evaluation of Radiological Emergency Response Plans and Preparedness in Support of Nuclear Power Plants (NUREG-0654/ FEMA-Rep.-1, 1980)*. Washington: Nuclear Regulatory Commission and Federal Emergency Management Agency.

United States Nuclear Regulatory Commission. 1981a. *NRR Project Manager Comments on Systematic Assessment of Licensee Performance* (May). Washington: Nuclear Regulatory Commission.

_____. 1981b. *Systematic Assessment of Licensing Performance NRR Performance Evaluation* (undated). Washington: Nuclear Regulatory Commission.

_____. 1982. *Systematic Assessment of Licensee Performance: Long Island Lighting Company* (April 26). Washington: Nuclear Regulatory Commission.

United States Office of Science and Technology, 1968. *Report on Considerations Affecting Steam Power Plant Site Selection*. Washington: U.S. Government Printing Office.

United States Office of Technology Assessment. 1984. *Nuclear Power in an Age of Uncertainty*. Washington: U.S. Government Printing Office.

United States Presidential Commission on Three Mile Island (The Kemeny Commission). 1979. *Report of the Commission*. Washington: U.S. Government Printing Office.

Urban Law Review. 1970. Nuclear Power--Emergency Planning Around Nuclear Power Plants. III: 299-313.

Wilmott, D., and M. Harrison. 1983. "The Shoreham Nuclear Power Station." In *Final Report of New York State Fact Finding Panel on Shoreham Nuclear Power Facility* (Sixty-six pages). Stony Brook, New York: New York State Fact Finding Panel.

Wilson, James. 1980. *The Politics of Regulation*. New York: Basic Books.

Wilson, Richard. 1987. "A Visit to Chernobyl." *Science*, 236: 1636-1640 (June 26).

APPENDIX

CHRONOLOGY OF SIGNIFICANT EVENTS IN THE SHOREHAM PROJECT

1965

April 13: LILCO announces plans to build the Shoreham nuclear plant, at a cost of $65-75 million.

1966

June: Brookhaven Town Board votes unanimously to issue zoning change to allow LILCO to build Shoreham.

August 16: LILCO's Board of Directors approves the selection of Stone and Wester for the environmental site survey at Shoreham.

1967

January 1: General Electric and Westinghouse abandon turnkey nuclear projects.

February 21: LILCO's Board of Directors authorizes the contract with General Electric to supply the turbine generator, nuclear steam supply system, and initial fuel load for Shoreham.

May 23: Edward Duffy, LILCO's executive vice-president, recommends the selection of Stone and Webster as architect/engineer and construction manager for the Shoreham project.

July 26: LILCO's Board of Directors approves the construction of Shoreham Unit 1, a 540 MW nuclear unit for commercial operation in the Spring of 1973 at a cost of $124 million.

October 9: LILCO announces plan for a nuclear facility at Lloyd Harbor.

1968

May 15: LILCO submits the Preliminary Safety Analysis Report (PSAR) for a 540 MW reactor at Shoreham to the Atomic Energy Commission.

October 1: LILCO vice-president Arthur Sugden recommends that the Long Island Lighting Company defer the Shoreham project in light of emerging concerns about nuclear plant construction.

October 30: LILCO Board Chairman Charles Touhy advises Board of Directors that recent Stone and Webster analysis has concluded that a 540 MW nuclear plant would be more expensive than an equivalent fossil fuel unit. Board decides tentatively to postpone Shoreham project, and authorizes a 380 MW fossil fuel unit to be built at Northport.

November 6: Stone and Webster Vice-President advises LILCO President Duffy that Stone and Webster had stopped work on the 540 MW unit, and thanks him for "authorization for the design and management of construction of an 820 MW unit" at Shoreham.

November 21: LILCO notifies the Atomic Energy Commission that it plans to construct an 820 MW unit at Shoreham with a commercial operation date of May 1975. The Atomic Energy Commission agreed to retain the same docket for the amended PSAR for the 800 MW Shoreham unit.

1969

February: LILCO issues a report saying that an 820 MW nuclear plant currently would have an economic advantage over an oil-fired plant.

March 26: LILCO Board of Directors authorizes the construction of an 820 MW unit at Shoreham at an estimated cost of $217 million, with a commercial operation date of May 1975.

April 21: LILCO files amended Preliminary Safety Analysis Report to reflect 820 MW Shoreham unit.

June: Arthur Wofford hired and named LILCO project manager for Shoreham.

July 1: Site clearing and grading begins at the Shoreham site.

September: LILCO increases estimated cost for Shoreham to $261 million, with projected commercial operation in May 1975.

November: Atomic Energy Commission proposes that all nuclear power plants receiving construction permits after April 1, 1970 be required to comply with ASME III standard, including B31.7 Piping Code.

November 8: Lloyd Harbor Study Group issues report strongly criticizing proposed Lloyd Harbor nuclear plant as being "unsafe, unreliable, and unclean." Subsequently LILCO abandons plans for construction of the facility.

December 15: The Advisory Committee on Reactor Safeguards issues a favorable letter on the issuance of the Shoreham construction permit.

1970

March 4: Stone and Webster recommends the use of the B31.7 code requirement for the design of piping systems at Shoreham.

August 24: Contract between Stone and Webster and LILCO for engineering and construction management services signed, effective retroactively to June 1, 1967.

September 1: Foundation work for turbine and reactor buildings begins.

September: LILCO increases the estimate of Shoreham's costs to $300 million with a projected commercial operation in May 1975.

September 21: Shoreham's construction permit hearings begin before a United States Atomic Energy Commission Atomic Safety Licensing Board.

1971

February 5: LILCO considers deferring the Shoreham project.

May: Atomic Energy Commission begins review of emergency core cooling systems in nuclear plants.

June 12: The Nuclear Regulatory Commission issues final regulations requiring plants with construction permits issued after January 1, 1971 to comply with ASME III, including B31.7 piping code.

July 23: United States Federal Court of Appeals issues *Calvert Cliffs* decision, requiring that the Atomic Energy Commission prepare environmental impact statements when considering licenses for nuclear plants.

August 3: LILCO instructs Stone and Webster to shut down engineering, design, and construction at the Shoreham project.

August: AEC regulatory staff issues an analysis saying that the proposed Shoreham emergency core cooling system is acceptably safe.

October 1: Stone and Webster halts Shoreham's construction.

November 4: Atomic Energy Commission Atomic Safety Licensing Board begins hearings on Shoreham's emergency core cooling system.

1972

January 31: LILCO increases its estimate of Shoreham's costs to $350 million, with an April 1977 commercial operation.

May 1: LILCO decides to reactivate engineering work for the Shoreham project.

June: Staff of Atomic Energy Commission issues Draft Environmental Statement for Shoreham concluding that construction permit is justified.

June: New York State Department of Environmental Conservation concludes that Shoreham can be built without violating State water quality standards.

September 27: LILCO's Board of Directors approves the selection of Stone and Webster for phase 1 of the proposed Jamesport nuclear project.

October 1: Construction and other work at the Shoreham site begins again after a one-year shutdown.

December 22: Atomic Energy Commission requests utilities to evaluate the effects in nuclear plants of high energy pipe breaks outside containment (PBOC), and provides criteria for preventing damage from such breaks.

1973

March 1: Thomas J. Burke becomes LILCO's project manager for Shoreham.

April 12: Atomic Energy Commission announces it is granting construction permit for Shoreham.

May 13: LILCO increases its estimate of Shoreham's cost to $560 million, with a commercial operation of July 1977.

June 14: First concrete is poured for the Shoreham reactor building.

October: First Arab oil embargo begins.

October 26: Atomic Safety and Licensing Appeal Board affirms the issuance of construction permit for Shoreham.

1974

February 28: Andrew W. Wofford reports that Stone and Webster cannot meet a target date of Spring 1978 for Shoreham's commercial operation.

March 26: LILCO's Nuclear Project Construction Group issues a "Self Analysis Report."

May 15: Charles R. Pierce becomes president of LILCO, and Andrew W. Wofford becomes LILCO vice-president of project management.

July 19: LILCO increases its estimate of Shoreham's costs from $506 million to $695 million, and changes target commercial operation from July 1977 to May 1978.

1975

January: United States Nuclear Regulatory Commission (NRC) assumes regulatory and licensing functions of Atomic Energy Commission (AEC).

January: To increase its direct contact with the project, LILCO moves the offices of its project manager Thomas J. Burke and the LILCO project team to the Shoreham site.

August 28: Final Safety Analysis Report and Environmental Report submitted to the NRC.

October: Suffolk County prepares the first emergency response plan for Shoreham in cooperation with LILCO.

November 7: The Nuclear Regulatory Commission rejects as incomplete LILCO's final safety analysis report.

December 8: The reactor pressure vessel is set at Shoreham.

1976

January 15: Piping System Erection Progress Report issued, citing numerous problems at the project.

January 23: Final Safety Analysis Report and the Environmental Report accepted for review by the NRC.

May 1: LILCO and Stone and Webster form UNICO (Unified Construction Organization) as the Shoreham construction manager with Stone and Webster in the lead role, and with direct participation by LILCO personnel.

June 30: LILCO increases its estimates of Shoreham's cost from $695 million to $969 million, and changes the target date of commercial operation from May 1978 to May 1979.

August 20: Piping erection task force is established with William J. Museler as chairman.

October 9: The first system at Shoreham is turned over to Start Up by construction.

December 7: Piping Erection Task Force Final Report issued.

December 20: Craft and Supervision Assistance Program (CASAP) final report issued.

1977

January 6: LILCO advises Stone and Webster that Stone and Webster would continue in lead construction manager role on a trial basis.

February 2: Nuclear Regulatory Commission rejects LILCO's approach to dealing with Pipe Break Outside Containment issues.

March 12: Stone and Webster advises LILCO that fuel load at Shoreham would be delayed between 15 and 28 months.

March 28: Arthur D. Little issues draft report to LILCO, discussing serious organizational problems at the Shoreham project.

July 5: Wilfred O. Uhl recommends that LILCO take over Shoreham construction management.

August 31: LILCO's Board of Directors approves the LILCO takeover of construction management at Shoreham.

September 12: LILCO assumes the lead role in UNICO and relieves Stone and Webster of construction management responsibilities.

September 14: LILCO agrees to comply with the intent of the Nuclear Regulatory Commission's rules on Pipe Break Outside Containment issues.

October: LILCO increases its estimate of Shoreham's cost to $1.223 billion, with a September 1980 commercial operation.

1978

January 1: Charles R. Pierce becomes Chairman of LILCO, and Wilfred O. Uhl becomes President of LILCO.

September 15: Joseph P. Novarro becomes LILCO's Shoreham project manager.

October 1978: LILCOs increases its estimate of Shoreham's cost to $1.337 billion, with a September 1980 commercial operation.

December 1978: Dr. William J. Catacosinos elected to membership on the LILCO Board of Directors.

1979

March 4: Three Mile Island accident occurs, resulting in a partial meltdown at the plant in Pennsylvania.

May 21: Public Service Commission begins investigation of the quality of management at the Shoreham project.

July 31: LILCO increases its estimate of Shoreham's cost from $1.240 to $1.581 billion, and changes expected commercial operation from September 1980 to May 1981.

December: Following problems of emergency response after the Three Mile Island Accident, the Nuclear Regulatory Commission proposes new emergency preparedness rules.

December: Suffolk County Executive John Klein and LILCO sign agreement updating County's responsibilities in emergency planning.

1980

July: LILCO increases its estimate of Shoreham's cost to $2.213 billion, with an expected November 1982 commercial operation.

August: The Nuclear Regulatory Commission adopts final new emergency preparedness rules, requiring more extensive emergency planning for nuclear power plants.

1981

July 1: William J. Muselar becomes LILCO's manager of construction and engineering at Shoreham.

July 24: LILCO increases its estimate of Shoreham's cost to $2.49 billion.

September: Suffolk County legislature approves a contract with LILCO to prepare an emergency plan for Shoreham.

1982

February: An attorney representing Suffolk County in the Shoreham operating license proceedings advises the County to cancel its emergency planning contract with LILCO. Suffolk advises LILCO that it is withdrawing from the contract.

May: LILCO forwards the initial Suffolk County-LILCO emergency plan for approval to New York State Disaster Preparedness Commission.

Suffolk County protests, calling the Suffolk County Planning Department document "resource material" only.

May: Suffolk County announces that it will prepare its own emergency response plan for Shoreham.

June 5: Nuclear Regulatory Commission Atomic Safety Licensing Board begins Shoreham operating license hearings.

September: Pre-operational testing of the diesel generators at Shoreham begins.

September 30: Torrey Pines Technology issues a report on Shoreham, recommending changes costing approximately $1 million but finding plant's construction and quality assurance program to be satisfactory from the standpoint of safety.

December: Suffolk County releases its draft emergency response plan, calling for planning for evacuation of up to twenty miles around Shoreham, rather than the ten required by Federal regulations, based on what it calls special circumstances on Long Island.

1983

January: Suffolk County conducts hearings on proposed emergency response plan for Shoreham.

February 16: Suffolk County Executive Peter Cohalan says that it is not possible to have adequate emergency planning for Shoreham, concluding that Shoreham therefore should never operate.

February 17: Governor Mario Cuomo orders the State not to approve any emergency plan not approved by Suffolk County. Hours after the Governor's executive order, the Suffolk County legislature votes 15-1 to cease emergency planning for Shoreham on grounds that effective emergency planning for Shoreham is not possible.

February 23: Suffolk County asks Nuclear Regulatory Commission to halt Shoreham's operating license hearings on grounds that the County had withdrawn from emergency planning, making it impossible for the NRC to grant LILCO an operating license for the plant.

April 19: Governor Cuomo proposes creation of a "Fact Finding Panel" for Shoreham, to be chaired by State University of New York at Stony Brook President John Marburger.

May: LILCO increases its estimate of Shoreham's costs to $3.3 billion, with expected commercial operation in April 1984.

May 12: Nuclear Regulatory Commission agrees to consider LILCO's emergency response plan, over Suffolk County's objections.

June: Consulting firm of Daverman and Associates reports to Suffolk County that acquiring LILCO's assets would cost Suffolk County about $1.4 billion, and would save Suffolk ratepayers between four and thirty percent on electricity bills. Suffolk County Budget Review Office subsequently criticizes report as being based on assumptions "whose validity remains uncertain."

June 8: LILCO files a motion with the NRC for a low power license for Shoreham.

August 12: Crankshaft on one of Shoreham's emergency diesel generators breaks.

October 1: Marburger Fact Finding Panel on Shoreham meets in executive session to consider conclusions for its Final Report to the Governor.

October 5: Federal Emergency Management Agency accepts Nuclear Regulatory Commission's referral of LILCO's emergency plan.

October 24: Stone and Webster recommends that LILCO purchase 3 Colt diesel generators to replace the Delaval diesel generators.

November: LILCO orders diesel generators from Colt Industries.

December: Final Report of Marburger Panel is released. Panel concludes that Suffolk County's decision to withdraw from emergency planning is not so unreasonable as to warrant intervention by Governor Cuomo to overturn decision.

1984

January 10: LILCO announces it will withhold property tax payments to Suffolk County on grounds that County was overassessing the Shoreham plant.

January 30: Charles R. Pierce resigns as LILCO chairman, and is replaced by Dr. William J. Catacosinos.

February 10: The Department of Public Service staff files its report on LILCO's management of Shoreham's construction. The report recommends that \$1.9 billion of Shoreham's \$4.2 billion cost to that point be disallowed because the cost resulted from poor management.

March 28: LILCO accepts the resignations of Wilfred O. Uhl, president, Thomas H. O'Brien and Charles J. Davis, senior vice-presidents, and Andrew W. Wofford, vice-president.

May: Suffolk County Legislature's Budget Review Office concludes that property taxes would need to rise up to sixteen percent if Shoreham was not taxed as an operating unit.

June: Suffolk County faces budget shortfall of \$200 million in its budget of \$800 million, a budgetary shortfall aggravated by LILCO's witholding of property taxes for Shoreham.

October: Atomic Safety Licensing Board issues low-power testing license for Shoreham.

October 23: Ten members of Suffolk County legislature introduce resolution to establish a public electric utility in Suffolk County; the resolution would prohibit the utility from owning or operating a nuclear plant.

November: Suffolk County's bankers raise concerns about the effects of LILCO's withholding of property taxes.

November: Federal Energy Secretary Donald Hodel argues to Suffolk County Executive Peter Cohalan that, based on data from the Three Mile Island accident, it might be possible to reduce the planning area for evacuation around a nuclear plant from ten miles to two miles.

1985

May 31: Peter Cohalan orders Suffolk County to cooperate with LILCO in testing emergency response plan for Shoreham. LILCO announces that it will cease withholding property taxes for Shoreham, and will pay $131 million in back taxes to Suffolk County. Suffolk County Legislature and Governor Cuomo oppose Cohalan's executive order.

June 10: Responding to an appeal by the Suffolk County Legislature, a New York State judge nullifies Cohalan's executive order.

June 20: Nuclear Regulatory Commission asks the Federal Emergency Management Agency (FEMA) to conduct exercise of LILCO's emergency response plan.

August: Shoreham begins low-power testing.

September: Hurricane Gloria hits Long Island. LILCO is criticized as parts of Long Island remain without power for eleven days. Chairman William Catacosinos is criticized for not returning from vacation to attend to hurricane crisis.

November 13: Chairs of the Senate and House Committees in charge of Federal Emergency Management Agency's appropriations write to FEMA saying that it should not allow state and local opposition to nuclear plants to undermine emergency planning efforts.

November 14: FEMA confirms that it will conduct test of LILCO's emergency plan.

December 16: The Public Service Commission announces that it will disallow $1.395 billion of Shoreham's costs as imprudently incurred, about one-third of the plant's cost to that point.

1986

January: New York State Public Service Commission grants LILCO $68.7 million rate increase, as recommended by its staff. Subsequently LILCO receives extension from banks on payment of loans. Governor Cuomo comments that "If I could start all over with the Public Service Commission, without confirmation, would this be the same Public Service Commission? No."

January: Republican State senators from Long Island endorse concept of a public power authority to replace LILCO on Long Island.

January 11: Suffolk County legislature passes a bill prohibiting exercise of emergency plan in the County without approval by the Legislature.

February 10: Federal judge bars Suffolk County from enforcing law preventing emergency planning exercise for Shoreham.

February 13: FEMA, the NRC, and LILCO conduct test of LILCO's emergency response plan.

April 14: FEMA Regional Director Frank Petrone resigns over wording of Post-Exercise Assessment. Petrone contends that the Assessment should specify that FEMA could not provide reasonable assurance that the plan would protect public safety because the local authorities did not participate in the exercise.

April 18: FEMA evaluates LILCO's test of emergency plan relatively favorably, but does identify note some deficiencies requiring corrections. Notes that emergency plan approval process normally requires more than one exercise.

April 26: Soviet nuclear reactor at Chernobyl explodes, spewing radioactive cloud over Europe which eventually reaches United States. Forty nine thousand people around Chernobyl are evacuated; thirty one die shortly, with more deaths expected from exposure to radiation.

May: Suffolk County Legislature votes to sue LILCO under the Federal Racketeer Influenced and Corrupt Organizations law, charging that LILCO had pursued rate increases from the New York Public Service Commission using arguments it knew to be false.

June: Nassau County Board of Supervisors refuses to allow LILCO to use Veterans Memorial Coliseum as emergency reception center. Nassau Chapter of American Red Cross tells the Atomic Safety Licensing Board that it cannot find school districts or other areas willing to provide transportation or shelter to evacuees.

June: Two panels, one for Suffolk County and one for New York State, evaluate a public takeover of LILCO relatively favorably.

July: New York State Senate (35-26) and Assembly (80-63) agree on bill establishing the Long Island Power Authority (LIPA), with an effective date of January 15, 1987. LIPA is empowered to undertake a friendly or hostile takeover of LILCO if it determines that a public power system would reduce electricity rate increases on Long Island.

September: Radio station WALK resigns as primary broadcaster of emergency messages in LILCO's emergency response plan.

1987

January: Governor Cuomo appoints five members to the Long Island Power Authority, with four additional representatives appointed by the leaders of the New York Legislature.

March 6: After discussions in February, the Nuclear Regulatory Commission formally proposes that it be allowed to *presume* that state and local governments would assist in any actual nuclear emergency; and that it be allowed evaluate emergency response plans, including those submitted by utilities, with that presumption.

August 5: House of Representatives rejects by a vote of 267 to 160 an amendment to the Nuclear Regulatory Commission's appropriation which would bar NRC approval of revised emergency planning procedures.

August: Long Island towns of Hempstead and North Hempstead declare that LILCO's decontamination trailers, an important part of its emergency response plan, would violate local zoning ordinances.

October 29: Nuclear Regulatory Commission votes unanimously to adopt revision of emergency planning rules, allowing it to presume that state and local governments will assist in any actual emergency.

November: New York State Public Service Commission, with five new members and a new Chair appointed by Governor Cuomo in previous year, refuses to grant LILCO a rate increase until Shoreham situation is resolved.

December 30: New York State Supreme Court upholds Public Service Commission's disallowance of excess engineering costs, construction labor costs, and diesel generator repair and replacement costs, but directs Public Service Commission to justify disallowance of delay costs in more detail.

1988

January: Four members of Long Island Power Authority object to the State's negotiations with LILCO--involving other members of LIPA--over a possible settlement of Shoreham in exchange for LILCO's survival as a private utility. The objecting members contend that LIPA's proposed takeover of LILCO is being stalled to accomodate the negotiations.

February: The Nuclear Regulatory Commission and the Federal Emergency Management Agency schedule another test of LILCO's emergency response plan for June 1988.

March 28: The Long Island Power Authority approves an initial "friendly" offer to purchase LILCO's preferred and common stock, and thus assume control over the company.

March 31: Chief State negotiators Vincent Tese and LILCO Chairman William Catacosinos meet in what is later called the pivotal session of the negotiations between the state and LILCO. Tese and Catacosinos agree on the outline of a settlement in which LILCO would abandon Shoreham, and the State would provide enough assistance to allow LILCO to return to financial health.

April: LILCO rejects the Long Island Power Authority's first "friendly" offer to purchase its stock. LIPA raises its bid, but also undertakes steps necessary to initiate a hostile takeover of LILCO.

April: The outlines of a proposed negotiated settlement between LILCO and the State, involving LILCO's selling of Shoreham and financial aid from the State, become publicly known.

May 25: LILCO and the Governor's Office agree to the settlement.

June: The Public Service Commission begins hearings on the proposed settlement, which it must approve. Conflicting studies of the costs of the settlement versus a public takeover of LILCO are presented.

June 22: The Long Island Power Authority votes 6-0 to accept the settlement, with three of Shoreham's opponents choosing not to vote or abstaining.

July: A *Newsday* poll reports that the public on Long Island divides 44-42 percent on whether they favor or oppose the settlement. The majority of those opposing the settlement do so because of the rate increases the settlement guarantees to LILCO.

July 16: The New York State legislature recesses for at least two weeks without having approved the settlement. Strong political opposition to the settlement by Long Island legislators begins to emerge.

September 1: Administrative Law Judge Frank Robinson, overseeing the Public Service Commission hearings on the settlement, concludes that the desirable option is to open and operate Shoreham. He also states that, if the Public Service Commission is willing to approve the abandonment of Shoreham, "the settlement is not otherwise objectionable; its rate arrangements are reasonable, and there would be no sound alternative scenarios."

September 2: The Federal Emergency Management Agency concludes that LILCO's emergency response plan would provide reasonable assurance of public safety in the event of an accident at Shoreham, even though state and local governments did not participate in the plan's exercise.

September 6: A Federal Appeals Court in Boston upholds the Nuclear Regulatory Commission's revision of its emergency planning regulations, allowing the NRC to presume that state and local governments will participate in an actual emergency response even though they have not participated in plan exercises.

September 6: Judge Jack Weinstein tells Suffolk County that it can represent only itself, and not LILCO's ratepayers, in its lawsuit of LILCO under the Racketeer Influenced and Corrupt Organization law. It is recognized, however, that ratepayers could subsequently collect damages if the suit is successful.

September 15: The Public Service Commission votes 6-1 to accept the Shoreham settlement negotiated by the State and LILCO.

September 23: An Atomic Safety Licensing Board (ASLB), in a 2-1 decision, authorizes the Nuclear Regulatory Commission to issue an operating license for Shoreham, and throws New York State, Suffolk County, and the town of Southampton out as intervenors in the case.

October: U.S. Secretary of Energy John Herrington writes to NRC Chairman Lando Zech asking the NRC to "move expeditiously" to license Shoreham.

October 3: The trial begins in Suffolk County's lawsuit of LILCO under the Racketeer Influenced and Corrupt Organizations Law.

October 7: A Nuclear Regulatory Commission Appeals Board overturns the September 23 decision of the ASLB, saying that the ASLB did not have the right to dismiss the State and County from the entire Shoreham proceeding.

October 30: A *Newsday* poll reports that 50 percent of Long Islanders oppose the State-LILCO settlement, and 31 percent favor it, largely because of the rate increases granted to LILCO.

November 4: LILCO's shareholders vote overwhelmingly to accept the settlement. LILCO Chairman Catacosinos gives the Legislature until December 1 to approve the settlement.

November 7: Governor Cuomo's political committee reports its own poll, concluding that 49 percent of Long Islanders favor the settlement and 26 percent oppose it. The question in the committee's poll includes the statement that "The New York State Public Service Commission has determined that the electric rates if the settlement is approved would be less than if the settlement is not approved and the Shoreham nuclear power plant is allowed to operate."

November 18: President Ronald Reagan issues an executive order specifying that the Federal government itself will provide emergency reception and medical centers, communication services, and other facilities if state and local governments fail to do so in an emergency at a nuclear plant.

November 29: In Suffolk County's RICO suit of LILCO, Judge Jack Weinstein suggests that the County's arguments and evidence are weak, but that he would not take the case away from the jury.

December 1: Legislature fails to approve the settlement by the December 1 deadline imposed by LILCO. However, it is recognized generally that negotiations may continue.

December 5: In the RICO lawsuit of LILCO, a six-person jury finds LILCO guilty of lying to Public Service Commission about Shoreham's progress in order to obtain rate increases.

December 16: Judge Jack Weinstein suggests that LILCO, Suffolk County, and the ratepayers settle the RICO lawsuit prior to his final action in the case or any appeals.

December 29: The New York Public Service Commission votes 6-1 to not grant any additional rate increases to LILCO until the basic terms of the original settlement are agreed to by the key parties, including LILCO, the Public Service Commission, the Long Island Power Authority, and the New York Power Authority.

1989

January 22: Kenneth Feinberg, a mediator proposed by Judge Weinstein, proposes outlines of a settlement of the RICO lawsuit to Suffolk County, LILCO, and attorney appointed by Judge Weinstein to represent ratepayers.

February 11: Judge Weinstein dismisses Suffolk County's lawsuit of LILCO under the Racketeer Influenced and Corrupt Organization law, concluding that the RICO law "cannot, and should not, be applied in a case such as this to permit a federal jury in a civil case to second guess the rate-making authority of the state." Although dismissing the case, Weinstein urges that the parties to settle the lawsuit before any further appeals are made.

February 11: Kenneth Feinberg proposes a RICO settlement to Suffolk County, LILCO, and ratepayers, providing $390 million in rate relief over ten years.

February 14: Judith Vladeck, attorney representing ratepayers, approves the settlement proposed by Feinberg.

February 15: Public Service Commission grants LILCO a rate increase, but conditions the increase on LILCO agreeing to not operate Shoreham even if granted an operating license. The PSC approves the increase 7-0, but the conditions for the increase only by a vote of 4-3.

February 27: Suffolk County legislature votes unanimously to appeal Judge Weinstein's dismissal of the RICO case, and refuses to settle the RICO lawsuit with LILCO.

February 28: Governor Cuomo announces that he and the Long Island Lighting Company have agreed on a new settlement to close Shoreham. Under the new settlement LILCO will sell Shoreham to the Long Island

Power Authority for $1, which will contract with the New York Power Authority for Shoreham's decommissioning. LILCO obtains a commitment from the State that it will make good faith efforts to return the utility to financial health. The new settlement will not require Legislative approval, and will leave final approval of rate increases solely to Public the Service Commission.

March 3: The Nuclear Regulatory Commission throws out Suffolk County and New York State as intervenors in the Shoreham proceeding, saying that the governments had acted in "bad faith" in "a willful attempt to obstruct the Commission's proceedings."

March 22: Judge Weinstein approves final settlement of the RICO suit for LILCO and its ratepayers, providing $390 million in rate relief, $10 million in attorneys' fees, and funding for a Citizens Advisory Panel to oversee implementation of the agreement. Suffolk County's legislators reaffirm their opposition to the settlement of the RICO suit, and announce they will press an appeal of Weinstein's dismissal of the case.

April 5: New York Public Service Commission approves new State-LILCO settlement by a vote of 6-1, and approves a ten-year plan for annual rate increases of between four and five percent for LILCO.

April 12: New York Power Authority votes 3-2 to arrange decommissioning for Shoreham and provide new power plants on Long Island if asked to do so as provided under the settlement.

April 12: James Watkins, the United States Secretary of Energy, calls the New York-LILCO settlement of Shoreham "one of the most foolish deals in the nation's history." Deputy Energy Secretary Henson Moore says that Department of Energy will "throw up every roadblock we can to it, and if we have to create some, we'll do that."

April 13: The Long Island Power Authority votes 8-1 to approve the new settlement.

April 13: United States Senate Committee on Energy and Natural Resources holds hearing on electricity supply in the Northeast which is generally critical of the New York-LILCO settlement.

April 14: LILCO Board of Directors votes unanimously to approve the new settlement.

April 17: Nuclear Regulatory Commission staff issues formal finding that Shoreham is acceptably safe and that Long Island's geography or other features do not "render emergency planning at Shoreham fundamentally more difficult than at other sites."

April 20: Nuclear Regulatory Commission votes to grant LILCO full commercial operating license for Shoreham.

June 28: LILCO's stockholders vote overwhelmingly to approve the New York-LILCO settlement leading to abandonment of Shoreham in exchange for a package of financial aid from the State.

June 29: U.S. House Committee on Energy and Commerce passes, 23-19, an amendment to NRC authorization bill sponsored by Representative Donald Ritter (R-PA) which prohibits the NRC from transferring LILCO's Shoreham license to the Long Island Power Authority.

July: U.S. Senate Committee on Energy and Natural Resources releases study by staff economist Karl Hausker criticizing New York's economic evaluation of settlement as being based on a "deeply flawed...obviously biased" study by the New York Public Service Commission.

July 18: House Speaker Thomas Foley delays full House vote on NRC authorization bill including the Ritter amendment.

August 15: President Bush, Deputy Energy Secretary Henson Moore, and Governor Cuomo meet on the Shoreham issue.

September 1: Federal government requests to join a suit in New York State which would block implementation of the settlement.

November 14: President Bush affirms administration policy that it would be better to operate Shoreham than to decommission the plant.

November 16: House debates NRC authorization including Ritter amendment. House votes 405-11 to delay final consideration of bill after Interior Committee Chairman Morris Udall says that Ritter amendment might constitute Legislative "taking" of LILCO's property, requiring Congress to compensate LILCO for the full cost of the plant.

1990

January 3: Governor Mario Cuomo's State of the State address calls for the possible conversion of Shoreham to a gas-fired facility.

INDEX

Technology, Risk, and Society

An International Series in Risk Analysis

1. J.D. Bentkover, V.T. Covello and J. Mumpower (eds.): *Benefits Assessment.* The State of the Art. 1986 ISBN 90-277-2022-3
2. M.W. Merkhofer: *Decision Science and Social Risk Management.* A Comparative Evaluation of Cost-Benefit Analysis, Decision Analysis, and other Formal Decision-Aiding Approaches. 1987 ISBN 90-277-2275-7
3. B.B. Johnson and V.T. Covello (eds.): *The Social and Cultural Construction of Risk.* Essays on Risk Selection and Perception. 1987 ISBN 1-55608-033-6
4. R.E. Kasperson and P.J.M. Stallen (eds.): *Communicating Risks to the Public.* International Perspectives. 1990 ISBN 0-7923-0601-5
5. D.P. McCaffrey: *The Politics of Nuclear Power.* A History of the Shoreham Nuclear Power Plant. 1991 ISBN 0-7923-1035-7

Kluwer Academic Publishers – Dordrecht / Boston / London